Indian Film

Erik Barnouw and S. Krishnaswamy

COURTESY CINÉMATHÈQUE FRANÇAISE

Indian Film

Second Edition

New York • Oxford • New Delhi
OXFORD UNIVERSITY PRESS
1980

First published, 1963, by Columbia University Press,
New York and London

Library of Congress Cataloging in Publication Data

Barnouw, Erik, 1908-
Indian film.

Bibliography: p.
Includes index.
1. Moving-pictures—India. I. Krishnaswamy,
Subrahmanyam, joint author. II. Title.
PN1993.5.I8B3 1980 791.43'0954 79-9358
ISBN 0-19-502682-9 ISBN 0-19-502683-7 pbk.

ON TITLE PAGES: PRELIMINARY SKETCHES BY SATYAJIT RAY FOR *Pather Panchali* (SONG OF THE ROAD)

Printed in the United States of America

For Dorothy and Mohana

Prologue

Many reasons account for the continuing and increasing interest in Indian films—not only among film makers but also among students of mass communication, political science, history, social psychology, musicology, dance, and related arts. Among reasons one might cite:

the astonishing output of the Indian film industry, for many years exceeding that of any other in the production of theatrical films;

its extraordinary hold over Indian audiences, exceeding that of any other medium;

its structure, featuring a huge public-sector documentary industry meshed with a private-sector fiction industry;

its persistent devotion to a song-and-dance tradition inherited from ancient Sanskrit drama;

its development of a parallel avant-garde that has earned world recognition and, occasionally, foreign exchange;

its fragmented financial structure, sometimes involving the thorny problem of "black money";

its maharajah-like stars, whose popularity sometimes propels them into politics;

its Indian music "tainted" by Western rhythms and instrumentation;

its multilingual enterprise, involving more than a dozen Indian tongues.

It is an industry of sharp contrasts, with work ranging from the noblest to the most preposterous, from the most hedonistic to the most devotional, from the most jovial to the most despairing. The contrasts stem from those of Indian society.

Indian films have been described as "purely escapist," but this is a judgment far beside the mark. The films serve a country marked by deep tensions—between wealth and poverty, old and new, hope and fear. The tensions are the basic material of the medium. Its images and sounds are only the means for playing on those tensions—sometimes subtly, sometimes powerfully.

Even in colonial days, films played a nationalistic role that sometimes baffled and eluded British censors. Throughout its history the Indian film world has touched on struggles between tradition and modernization. In doing so, it may to some extent have defused those conflicts, providing a dreamland substitute for action. Yet it has also brought continued pressure on them, often foreshadowing and furthering social change.

All these factors were noted in the first edition of *Indian Film,* which was largely researched during 1960-62 and appeared in 1963, as the industry celebrated its fiftieth anniversary of feature production. The celebration stimulated new interest in Indian film history, to which the edition may have contributed. In researching it, the authors had virtually excavated new territory, sometimes to the puzzled wonderment of film people themselves. There were no archives. Few memorabilia had been treasured or even preserved. Film pioneers had been forgotten; some lived in destitution. Many films had been lost. Yesterday's films seemed to have little meaning for film makers, and were assumed to be irrelevant. Only today's hit counted. That film history should be studied as a scholarly subject appeared also to surprise scholars. Why not study cave temples, they asked—or Bharata Natyam—or classical music? But Indian *film* . . . ?

In researching this second, updated edition after a lapse of almost two decades, the authors have been gratified to find the atmosphere changed. A new generation of film workers has grown up, informed and stimulated by new factors—a film society movement burgeoning throughout the 1960s; the Film and Television Institute of India launched in 1961; the National Film Archive of India founded in 1964; the International Film Festivals, which became annual events in the 1970s; and the various national and state awards established during these years. All these had given film makers a sense of the social role of film, and of themselves as part of an evolving and meaningful enterprise. While profits made by major hits have encouraged fidelity to the status quo, new awareness has also spurred an exploratory spirit. The authors have been pleased to note these diverse pressures, and the resulting fruitful ferment.

We are grateful to the numerous producers, directors, actors, writers, cameramen, composers, scene designers, scholars, film society leaders, public servants, and others who have aided the research for *Indian Film.* A list of those interviewed, both for the first edition and the present updated version, will be found in the Reference section of this book. The authors also acknowledge their debt to the Fulbright Commission, which made possible research for the first edition, and to the binational Indo-U.S. Subcommission on Education and Culture, which underwrote research for the new edition.

Among those whose help in the updating work has been beyond the call of international cooperation we must mention Jagat Murari and Satish Bahadur of the Film and Television Institute of India; and P. K. Nair and Devi Kane of the National Film Archive of India. For their special help to the original research we must again thank A. R. Baji, V. Balakrishnan, Manujendra Bhanja, R. V. Ishwar, S. Krishnamurti, I. K. Menon, Subodh Mukherji, and Veeranki Rama Rao.

ERIK BARNOUW
Washington

S. KRISHNASWAMY
Madras

Contents

Illustrations

Note

The *anna* was 1/16 of a *rupee*. In 1957 India began converting to a decimal system in which a *rupee* became 100 *paise*. The *anna* persisted for a time in popular language. For rule-of-thumb calculation, readers may find the following useful. In terms of American currency, the *rupee* stood at 30-40¢ during early periods covered in this book. After independence the rupee was set at 21¢, but in 1966 was officially devalued, subsequently standing at around 13¢.

Indian Film

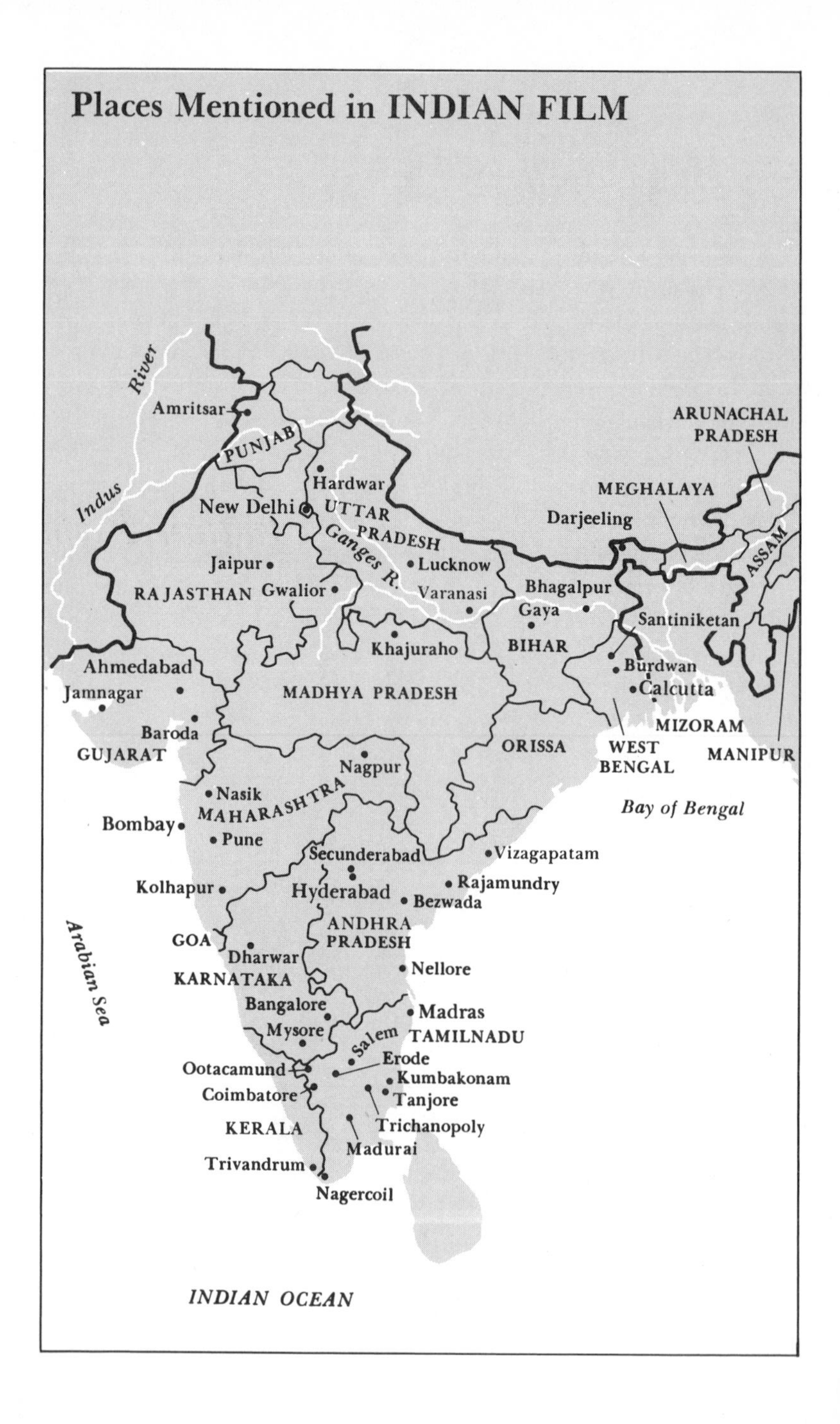
Places Mentioned in INDIAN FILM
Indus River
Amritsar
PUNJAB
Hardwar
New Delhi
UTTAR PRADESH
Ganges R.
Darjeeling
ARUNACHAL PRADESH
MEGHALAYA
ASSAM
Jaipur
Lucknow
RAJASTHAN
Gwalior
Varanasi
Bhagalpur
Gaya
Santiniketan
Khajuraho
BIHAR
Ahmedabad
Burdwan
Jamnagar
Calcutta
MADHYA PRADESH
Baroda
MIZORAM
GUJARAT
ORISSA
WEST BENGAL
MANIPUR
Nagpur
Nasik
MAHARASHTRA
Bay of Bengal
Bombay
Pune
Secunderabad
Vizagapatam
Rajamundry
Kolhapur
Hyderabad
Bezwada
Arabian Sea
ANDHRA PRADESH
GOA
Dharwar
Nellore
KARNATAKA
Bangalore
Madras
Mysore
Salem
TAMILNADU
Erode
Ootacamund
Kumbakonam
Coimbatore
Tanjore
Trichanopoly
KERALA
Madurai
Trivandrum
Nagercoil
INDIAN OCEAN

Beginnings

Film history began in India on July 7, 1896, when a *Times of India* advertisement invited Bombay residents to witness "the marvel of the century, the wonder of the world" at Watson's Hotel that same day. The attraction was described as "living photographic pictures in life-sized reproductions, by Messrs. Lumière Brothers." It was called the *cinématographe* and would be shown at 6, 7, 9 and 10 P.M.

Very likely few Bombay residents had previous intimations of the existence of the *cinématographe*. It was new, having been unveiled by the Lumières at a Paris café only a few months earlier—December 28, 1895. The advertisement may have seemed the hyperbole expected of showmen.

The hotel demonstrations, each consisting of many short items, began quietly but stirred enough excitement to prompt additional showings at the Novelty Theatre. These started on July 14. They were announced as "for three days only" but were held over, although the monsoon had begun. After two weeks of these showings the *Times of India* carried a further announcement:

> At the desire of a large number of Bombay residents who have flocked recently in spite of bad weather to see the Kinematograph, the patentee has obtained a fresh lease of the Novelty Theatre for a few more nights.[1]

But a few days later, the show was again held over.

A notable aspect of these events is that they paralleled similar, concurrent unveilings in cities around the world. The day on which the *cinématographe* was revealed to Bombay was also the day on which another Lumière expedition was showing the wonder to the Tsar of Russia in St. Petersburg. Other traveling missions, some representing rival entrepreneurs, were at about the same time introducing the miracle of living pictures to audiences in China, Australia, South Africa, and elsewhere. Most European countries had seen it earlier, but only by weeks or months. New York audi-

[1] *Times of India,* July 27, 1896.

ences had been introduced to it in April. Film history was, in short, erupting almost simultaneously on every continent—and stirring a strange new fever.[2]

Many factors were behind the extraordinary diffusion. One was that the Lumière brothers, before exposing their invention publicly, had planned a world-wide campaign to reap its first benefits. They had reasons for doing this. The Edison peepshow *kinetoscope,* which had appeared in 1894 and from which the Lumière brothers had derived the basic technology for images in motion, had not been patented internationally.[3] Many inventors therefore felt free to rush toward the next step, which was fairly obvious and was discussed in various journals—a *projecting kinetoscope,* linking the device to a magic lantern to serve audiences rather than individuals. It was a free-for-all race involving many experimenters.

The Lumière brothers—mainly Louis—devised their remarkable *cinématographe* early in 1895, yet held back its public debut. The several dozen short *cinématographe* films made during 1895 were all shot with the first machine by the brothers themselves, while they carefully guarded its secrets. By the end of the year they had twenty-five machines ready and more in production, and had trained *opérateurs* to use them. These were the ingredients of the whirlwind drive that was to follow the debut.[4]

The *cinématographe* lent itself spectacularly well to such a campaign. Compact and portable, it could—with slight adjustments—serve as camera *or* projector *or* printing machine. Thus an *opérateur* with a *cinématographe* was a working unit. He could be sent to a foreign metropolis to reveal the miracle with showings of the Lumière films, earning quick revenue; and he could meanwhile shoot new material, develop it in a hotel room at night, and within days announce a "change of program" that could include local scenes, to produce new astonishment. These could also be sent back to France to enrich the Lumière catalogue for subsequent expedi-

[2] For the wide travels of the Lumière expeditions see Sadoul, *Louis Lumière;* and Deslandes and Richard, *Histoire Comparée du Cinéma*. For the China film debut, see Leyda, *Dianying,* p. 1. The first film showings in South Africa and Australia were by magician Carl Hertz in the spring and summer of 1896. He later took his film show to India. See Barnouw, "The Magician and the Movies," *American Film,* April, May, 1978.
[3] Ramsaye, *A Million and One Nights,* p. 76.
[4] Sadoul, *Louis Lumière,* pp. 55–76.

tions. This was the agenda for action as the "marvel of the century" was publicly launched at the Grand Café in Paris.

Begun in a modest setting, the showings soon generated an atmosphere of hysteria. Within weeks Lumière film shows were running simultaneously at four Paris locations, and rival inventors were making and selling equipment said to produce similar miracles. Among those rushing to buy were touring conjurors eager to include the marvel in their "magic" acts.[5] Meanwhile Lumière *opérateurs,* with instructions to withhold the secrets of their equipment from everyone including kings and beautiful women,[6] began crisscrossing the globe—Europe, America, Africa, Asia, Australia. Other traveling showmen were close behind, sometimes ahead, as the *cinématographe* expeditions sped to major world centers.

The Lumière envoy who won the race to Bombay was Maurice Sestier. En route to Australia, he would first proclaim the good news to India. In one aspect of his mission he apparently failed. An Australian portrait photographer, Walter Barnett, made Sestier's acquaintance during the Bombay stay and found him in a troubled state. Sestier had received a stiff rebuke from Lumière headquarters, to the effect that the films he had sent back were incompetent. This may explain why no Indian scenes were listed in the first Lumière catalogues, though dozens of other countries were represented.[7]

But in other respects the mission to Bombay followed the pattern, starting quietly but gaining momentum. Like many Lumière expeditions, it adjusted its procedures to local custom. Newspaper stories about the event resembled those appearing elsewhere, suggesting that the missions carried efficient publicity material. In its first review the *Times of India* commented:

[5] Robert W. Paul, a British instrument maker, was among those who had begun motion picture experiments concurrently with the Lumières. As a result of the show-business frenzy created by the advent of the *cinématographe,* Paul experienced a "phenomenal rush" for his equipment, "first from conjurors." It was from Paul that the magician Carl Hertz obtained equipment for his South African, Australian, and Indian tours, though he advertised his film shows as the "cinematograph." See Hertz, *A Modern Mystery Merchant,* and Paul, B.K.S. lecture, February 3, 1936.
[6] Leyda, *Kino,* p. 18.
[7] Sestier was more successful in Australia, where he shot *Melbourne Races,* which survives and is included in the Australian compilation film *The Pictures That Moved.* Baxter, *The Australian Cinema,* pp. 2–3.

NEW ADVERTISEMENTS.

THE MARVEL OF THE CENTURY.
THE WONDER OF THE WORLD.
LIVING PHOTOGRAPHIC PICTURES
IN
LIFE-SIZED REPRODUCTIONS
BY
MESSRS. LUMIERE BROTHERS.
CINEMATOGRAPHE.
A FEW EXHIBITIONS WILL BE GIVEN
AT
WATSON'S HOTEL
TO-NIGHT (7TH instant).
PROGRAMME will be as under:

1. Entry of Cinematographe.
2. Arrival of a Train.
3. The Sea Bath.
4. A Demolition.
5. Leaving the Factory.
6. Ladies and Soldiers on Wheels.

The Entertainment will take place at 6, 7, 9, and 10 p.m.

ADMISSION ONE RUPEE.

CONSIGNEES of Cargo, per Hansa Line Steamer "HOCHHEIMER," from BREMEN, HAMBURG and ANTWERP, are hereby informed of the ARRIVAL of the steamer and requested TO TAKE IMMEDIATE DELIVERY of their GOODS landed at the Victoria Dock.

The Company's Surveyor will attend every morning to survey damages, and all Claims in this respect must be adjusted before the Goods are removed from the Landing Sheds.

All Claims must be presented before the sailing of the Steamer, otherwise none will be entertained.

GLADE & Co.,
Agents, Hansa Line of Steamers.

Bombay, 6th July, 1896.

SHIPPERS and CONSIGNEES of CARGO, per Italian Steamer "GIAVA," Captain PIETRO PIVANDELLO, are informed of her ARRIVAL from GENOA on the 1st instant, and that she is loading and discharging in Victoria Dock, and all claims for short delivery, &c., must be sent in before the departure of the Steamer, otherwise they will not be admitted under any circumstances.

L. CROCCO.
F. MASSARI,
Agents.

Sassoon House, Elphinstone Circle.

NOTICE.—LOST, two BOND WARRANTS No. 86 and 644, for 75 packages marked S B H C M (in a diamond) $1,472-71 and S B H C M (in a diamond) $1,023-47. Delivery for same stopped. Finder rewarded on returning to HALLIDAY & Co., Sirdar's Buildings.

REQUIRED 200 to 300 Hindustani COOLIES for Coffee and Tea Estate Work, under agreement for a term of three years. Rs. 10 will be advanced, provided the man or Company produces security to firms to an equal amount. Wages will be paid monthly at liberal rates. For particulars apply to the Undersigned

ROBERT WARDROP,
Ambawella Estate, Ambawella, Ceylon.

FIRST INDIAN FILM ADVERTISEMENT, *Times of India*, JULY 7, 1896
COURTESY ASIATIC SOCIETY OF BOMBAY

The life-like manner in which the various views were portrayed on the screen by the aid of a powerful lantern, and the distinctness with which each action of moving bodies were [*sic*] brought out showed to what an advanced stage the art of photography and the magic lantern had been brought, something like seven or eight hundred photographs being thrown on the screen within the space of a minute. The views being of a varied character found much favour, the more crowded scenes being applauded by the audience.[8]

The shows were advertised regularly in the *Times of India*. The phrase "tonight entire change" appears in several advertisements.

By the end of July the showings had acquired two indigenous aspects. "Reserved boxes for Purdah Ladies and their Families" were announced late in July. And a broad scale of prices was introduced. For the first showing there had been a single admission price of one rupee. Now prices ranged from a low of four annas to a high of two rupees.[9] This wide price range was to remain a feature of film exhibition in India, important to its future growth and range of appeal.

In early August the drawing power of the attraction seems to have waned. The *Times of India* editorially rebuked "our Parsee friends" for not taking more interest in the unique event.[10] The addition of "selections of suitable music under the direction of S. Seymour Dove" does not seem to have helped, and August 15 was announced as "POSITIVELY THE LAST exhibition in Bombay." A performance of *The Pickpocket* by the Thespian Club—"Soldiers & Sailors Half Price to Back Seats"—was already advertised for the following week.[11]

The Lumières' sense of urgency was justified by events of the following months. In January, 1897, "Stewart's Vitograph" (*sic*) came to the Gaiety Theatre and apparently ran about a week.[12] In September the "Hughes Moto-Photoscope, the latest marvel in cinematographs," began showings at various locations including fairgrounds.[13] The following year brought a Professor Anderson and Mlle Blanche and their "Andersonoscopograph" exhibiting varied items.[14] While Bombay was receiving these, Calcutta, at this time

[8] *Times of India*, July 22, 1896. [9] *Ibid.*, July 27, 1896.
[10] *Ibid.*, August 5, 1896. [11] *Ibid.*, August 15, 1896. [12] *Ibid.*, January 4, 1897.
[13] *Ibid.*, September 15, 1897. [14] *Ibid.*, December 26, 1898.

the capital of British India, was also visited by various expeditions, including that of a Mr. Stevens who is said to have exhibited short items at the Star Theatre after stage performances.[15]

It seems clear, in spite of the four-anna seats and the attention to purdah ladies, that these early showings attracted mainly British residents, along with a few Indians "of the educated classes"—especially those who identified their interests with those of the British. At the same time, the impact on Indians who attended was crucially important. Among those who saw the Lumière exhibition was Harischandra Sakharam Bhatvadekar, a Maharashtrian. According to his obituary notice in a Bombay trade publication,[16] Bhatvadekar had opened a photographic studio in Bombay about 1880. In 1896 he was so "hypnotized" by the Lumière showing that he ordered a motion picture camera from London, at a price of 21 guineas—probably the first imported. When it arrived the following year he photographed a wrestling match at Bombay's Hanging Gardens, and sent the film to London for processing. He had meanwhile bought a projector and become an itinerant, open-air exhibitor of imported films. Among these, months later, he was able to show his own wrestling-match film. His second subject is said to have been the training of circus monkeys. More important was his coverage of an event of December, 1901. An Indian student at Cambridge, R. P. Paranjpye, had won special distinction in mathematics, and his return to India was an occasion for wild jubilation and garlanding. It was the sort of occasion that aroused nationalist emotions in Indian hearts, and at the same time was noted with prideful interest by the British. It thus received enormous attention and has won a place in some Indian film chronologies as "the first newsreel event."[17]

In 1903 the durbar that celebrated the coronation of Edward VII with oriental and occidental splendor was another event photographed and shown by Bhatvadekar. His work as pioneer exhibitor led to a career as manager of Bombay's Gaiety Theatre—later re-

[15] *Amrita,* December 29, 1961.

[16] *Indian Documentary,* Vol. IV, Nos. 3–4 (1958).

[17] See, for example, "Landmarks in Indian Film Story," in *Indian Talkie, 1931–56,* pp. 17–18. R. P. Paranjpye eventually became vice-chancellor of the University of Poona.

named the Capitol Cinema. He eventually gave up production for exhibition and, perhaps in consequence, died with "quite a fortune."[18] Not all his fellow pioneers were similarly blessed.

Bhatvadekar's career suggests the rapid pace of events. The traveling missions from Europe and America were quickly followed by importation of films, projectors, and other equipment. Some of the missions, in fact, functioned as sales agents. Among the purchasers, a number took up cinematography and began to turn out such items as *Poona Races '98* and *Train Arriving at Bombay Station*—both advertised in December, 1898.[19] The typical film showman of the time, as elsewhere in the world, was the photographer-exhibitor.

Films continued to turn up in theatres, sometimes as supplements to plays, concerts, or performances of magic. In Bombay, in 1898, Carl Hertz, "absolutely the world's greatest conjuror," offered film items in color along with his magic show. In Calcutta, Hiralal Sen, who purchased equipment in the same year,[20] photographed scenes from some of the plays at the Classic Theatre; such films were shown as added attractions after the stage performances. But the importance of such events was overshadowed, for the time being, by the eruption of outdoor cinema shows, in tents or in the open air.

Tent to palace

The showman generally equipped himself with films for two or three programs. Having exhausted the possibilities in one location, he moved elsewhere. Showings in parks and empty lots of big cities soon led to showings in smaller cities and towns and eventually to the rural "traveling cinemas" still important in India.

Jamjetji Framji Madan (1856–1923), member of a Parsi family that had moved from Bombay to Calcutta, became interested in the theatre at an early age. Calcutta, as well as other of the larger Indian cities, was experiencing a rebirth of theatre. This had begun

[18] *Indian Documentary,* Vol. IV, Nos. 3–4 (1958).
[19] *Times of India,* December 26, 1898. Bardèche and Brasillach, in *History of Motion Pictures,* p. 9, have commented: "Quite a number of trains arrived and departed in the early films."
[20] Chakrabartty, "Bengal's Claim to Pioneership," *Dipali,* April 8, 1939.

during the 1830s and had slowly gained momentum, although only among educated strata of Indian society. Madan started as prop boy at Calcutta's Corinthian Hall, later toured other cities of India as an actor, and eventually purchased the company in which he had started. Madan, along with various relatives, was involved in innumerable enterprises. He was an importer of liquors, foods, and pharmaceutical products, and dealt in insurance and real estate. Throughout life he combined such activities successfully with his theatrical interests.

In 1902, having purchased film equipment from an agent of Pathé Frères, he launched a "bioscope" show in a tent on the Maidan, the green in the heart of Calcutta.[21] This was the beginning of what was to turn into a film production-distribution-exhibition empire, a powerful factor for three decades, not only in India but also in Burma and Ceylon.

Another film magnate of later years, the venturesome Abdulally Esoofally (1884–1957), likewise began as a tent showman. From 1901 to 1907 he moved throughout southeast Asia, holding "bioscope" showings in Singapore, Sumatra, Java, Burma, Ceylon; from 1908 to 1914 he continued his cinema travels in India. His tent was 100 feet long and 50 feet wide, propped by four posts, and could hold a thousand people. The short items shown were purchased outright by Esoofally, according to the practice of the time, and were used till the prints wore out.

> I had to buy these bits at the rate of 6d. per foot and 40 or 50 pictures composed my full programme. The films, however small, provided a varied fare. They included comedy gags, operas, travel films, sports events, etc. The maximum length of those films ranged between 100 and 200 feet and only in 1908 I remember to have shown my biggest films—1,000 feet in length—in my traveling cinema. When I started my bioscope shows in Singapore in 1901, little documentary films I got from London helped me a lot in attracting people. A short documentary about Queen Victoria's funeral and another about the Boer War showing the British Commander-in-Chief Lord Roberts' triumphant entry into Pretoria against the forces of Paul Kruger, the President of the Transvaal Republic, proved wonderful draws. People

[21] The date is variously given. J. J. Madan, as managing director of Madan Theatres Ltd., gave 1902 as the starting date of Madan film enterprises, in testimony before a government inquiry in 1927. *Evidence,* II, 829.

who had merely heard or read some vague reports about the war were thrilled beyond description when they saw the famous figures of the Boer War in action.[22]

In 1914 Esoofally finally settled down by taking over, with a partner, the Alexandra Theatre in Bombay; in 1918 they built the Majestic Theatre, where they were later to premiere the first Indian talking feature.

The imported films shown by early traveling showmen came from many countries, but American and French films might come to them via London distributors. Many were well worn before being dumped on the Asian market. Worn prints remained, for decades, a problem plaguing Asian exhibitors.

Throughout the early years the length of imported films grew rapidly. In the catalogues of one English producer, distributor, and equipment manufacturer, James Williamson, films averaged 60–75 feet in length in 1899 but ran to 280 feet in 1902.[23] As films lengthened they acquired, or aimed at, more substantial content. A "grand cinematographic programme" at Bombay's Gaiety Theatre in the 1901 Christmas season offered *Life of Christ,* "showing Birth, Miracles, Trial, Sufferings, Crucifixion, Burial, Resurrection, and Ascension." It also included, in what must surely have been one of the least merry of holiday programs, *The Queen's Funeral Procession* and *Assassination of President McKinley.*[24]

During the following years Europe and America experienced an increase in productions based on literary classics. The era of the pretentious Film d'Art, often featuring stage stars, began in 1907. For several years products of this sort dominated European and American cinemas and also the cinemas establishing themselves on a permanent basis in the big cities of India. In Calcutta J. F. Madan built the Elphinstone Picture Palace, the first of many Madan film theatres, in 1907. During the 1910s he expanded steadily and by the end of the decade had thirty-seven theatres.[25] In Bombay, after 1910,

[22] "Half a Century in Exhibition Line: Shri Abdulally Recalls Bioscope Days," in *Indian Talkie, 1931–56,* pp. 121–22.
[23] Low and Manvell, *History of the British Film, 1896–1906,* p. 45.
[24] *Times of India,* December 23, 1901.
[25] *Evidence,* II, 844.

the rivalry among film theatres, as reflected in the growing size and fulsomeness of newspaper advertisements, grew intense.

Along with the stagy dramas, comics were now a booming attraction and would soon emerge from anonymity into stardom. A week in September, 1912, found the Imperial Cinema in Bombay showing *The God of the Sun* along with various Pathé items and "two screaming comics." The Alexandra Theatre had a two-hour show including five "ripping comics." The America-India, apparently the first theatre to install electric fans, offered *The Mystery of Edwin Drood, The Dance of Shiva,* and "three real good bits of fun." The Excelsior had an all-French program, while the Gaiety, "the Rendez-vous of the Elite of Bombay," was announcing a season of "London's latest successes by the Ambrosio, Lubin, Vitagraph, American Bioscope, Nordisk, Urban, Pathé and other film companies."[26]

Clearly the film scene in India, as in other countries, was at this time extremely international. France, headed by Pathé, was apparently the leading source, but products of the United States, Italy, England, Denmark, and Germany also competed for a share of the Indian market. To this complex struggle a new element was about to be added, and it came from a totally unexpected source.

Enter a shastri

Dhundiraj Govind Phalke, more generally known as Dadasaheb Phalke, was born of a priestly family at Trimbkeshwar, in the district of Nasik not far from Bombay, in 1870. Committed by birth to be a shastri, a learned man, he was trained for a career as Sanskrit scholar, in emulation of his father. But he early showed a feverish interest in painting, play acting, and magic. The family moved to Bombay when the father joined the teaching staff of Elphinstone College, and this made it possible for young Phalke, at the completion of high school, to study at the Sir J. J. School of Art, a large institution in Bombay. Here he received a grounding in various arts including photography. He had also, by now, become a skilled magician, which talent he later put to professional use. After further

[26] *Times of India,* September 14, 1912.

art training at the Kala Bhavan[27] in Baroda and a period as photographer for the governmental Archaeological Department, he was offered backing to start an Art Printing Press. He now settled down, to all appearances, to a life of fine printing. He was married, and raising a family.

His backers, to acquaint him with the latest printing processes, especially in color work, arranged for him to take a trip to Germany. The arrangement provided that Phalke must remain with the company at least a stipulated time after the journey, which he did. But he already knew, when he returned, that a printing career would not satisfy him. About 1910 he fell ill and for a time lost his eyesight. On the return of his vision, he had an experience that determined the course of his life.

At a Christmas cinema show he saw a *Life of Christ.* Before he got home, a determination had formed in him. He asked his wife to go with him to the next showing. Family tradition has it that there was no cash in the house, and money had to be borrowed from neighbors for transportation and cinema tickets. Meanwhile Phalke explained what was on his mind. He now knew what he must do with his life.[28]

As he had watched *Life of Christ* he had been thinking about the possibilities of a film on Lord Krishna, most beloved of Hindu deities. The rescue of the infant Krishna from deadly perils, the pranks of his boyhood, his many romantic involvements, his love of Radha, his later wisdom and exploits of valor, were already taking cinema form in the mind of Dadasaheb Phalke. Many currents of Phalke's life—priestly lineage, dramatic appetite, technical virtuosity—merged in this project. By the time the film show was over, his wife fully understood his plan and what it would involve, and she agreed. She became, in fact, a most important collaborator.

There were family councils, in which various relatives voiced disapproval. The recklessness of the plan, which involved giving up printing, appalled them. But in the end he went ahead. With help from some, and funds raised by mortgaging his life insurance, he engaged passage for England, to obtain equipment and guidance. He

[27] Now the art department of the University of Baroda.

[28] Interview, Neelakanth Phalke; and *Phalke: Commemoration Souvenir,* p. 105.

prepared for the trip by buying, at a Bombay bookstall, an *ABC of Cinematography*,[29] apparently the work of the British film pioneer Cecil Hepworth. In England Hepworth was one of the film makers visited by Phalke.

Early in 1912 Phalke returned to India with a Williamson camera, a Williamson perforator—film had to be perforated in the darkroom before use—developing and printing equipment, raw film for several months of work, and a collection of the latest film publications.

Phalke did not have enough funds for a major film so he began with an intermediate project—economical but at the same time expressive of his inquisitive spirit. He decided on a short film in time-lapse photography. In the Phalke home the precious camera, zealously guarded from the children by Mrs. Phalke—Saraswathi or "Kaki" Phalke—was mounted before a pot of earth. Dadasaheb Phalke worked out the mechanism for intermittent photography. Finally friends, including a prospective financier, were invited to see the result: a capsule history of the growth of a pea into a pea-laden plant. The audience was astounded, and Phalke got his financial backing.[30] His backer at this time was Nadkarni, Bombay dealer in photographic goods.[31]

Still postponing the crucial Krishna project, Phalke now decided on a slightly easier topic, likewise based on Indian mythology and judged by Phalke to have powerful appeal. The story was that of Harischandra, a king so devoted to truth and duty that for their sake he sacrificed everything including wealth, kingdom, wife, and child—but was rewarded in the end for his steadfastness. The story, from the *Mahabharata*, was known to every Indian via uncounted centuries of oral tradition.

The difficulties facing Dadasaheb Phalke must have seemed, at times, insuperable. Although theatre and its sister art, the dance, were supposed by the Hindus to have originated with the gods—Brahma himself had ordered the first dramatic performance—and

[29] Interview, Neelakanth Phalke.
[30] Suresh Phalke, "The Film Industry and Phalke," *Hindustan Standard*, February 24, 1961.
[31] Interview, Neelakanth Phalke.

PHALKE AT THE CAMERA

although theatre had had a golden age in the time of Kalidasa and had lived on into later eras in the courts of kings, it had in recent centuries become an outcast, and almost perished in dishonor. Alien invaders, in successive waves, had taken the citadels of power and conferred prestige on other arts, other values. Indian theatre and dance had lost their standing and become a domain of the degraded castes, the occupation of prostitutes. So strong had become the association between the performing arts and the prostitute that measures to combat prostitution seemed likely, for a time, to eradicate what was left of Indian drama, dance, and music.[32] The new theatrical tradition, at first inspired by European example, that was slowly gathering strength in Calcutta and a few other cities had not yet pushed beyond a privileged group.

For Phalke, all this simply meant that no decent Indian woman would think of acting in a film. In fact, Phalke knew he could

[32] See the discussion in Bowers, *Theatre in the East,* pp. 15–38.

not ask a decent woman to do so. He approached several prostitutes but none would consider the prospect. The invitation to appear on the cinema screen seemed to them an invitation to be publicly branded. Phalke had to find other solutions.

In a restaurant he watched a young man at work, a cook with slender features and hands. Phalke asked him what he was earning. Ten rupees a month, the young man said. Phalke offered him fifteen to work in Phalke's films. The young man, A. Salunke, thus joined the enterprise and played the heroine Taramati.[33] In a later Phalke film, *Lanka Dahan* (The Burning of Lanka), based on a portion of the *Ramayana,* Salunke was to play both hero and heroine. He was the beautiful Sita, held by the ten-headed monster Ravana on the island of Lanka; and he was also Rama on the mainland, organizing the invasion army of men and monkeys. The film, Phalke's greatest success, made Salunke at one time indisputably the most popular film actor and actress in India.

Rajah Harischandra (King Harischandra), the first Phalke film, seems to have been completed late in 1912. Produced just before the era dominated by the long feature, it was a step ahead of its time: a massive project, 3,700 feet long.[34] It opened at the Coronation Cinema in Bombay early in 1913 and was an overwhelming success. It was followed, during the next two decades, by more than a hundred other Phalke films, ranging from short items to ambitious features.

The audiences attending the Western films at the new cinema palaces paid little attention to Phalke. The English-language newspapers hardly noticed him, and Phalke did not advertise in them. He was reaching, almost at once, a different public. To his audience, *Rajah Harischandra* and its successors—*Bhasmasur Mohini* (The Legend of Bhasmasur), *Savitri* (Savitri), *Lanka Dahan* (The Burning of Lanka), *Krishna Janma* (The Birth of Krishna), and others—were like revelations. To them the inhabitants of the Western films

[33] Interview, Neelakanth Phalke.

[34] The length in minutes, in the days of handcranking, seems to have been unpredictable. Phalke once complained: "The usual projection speed should be about three to four thousand feet per hour but sometimes the projectors even show about eight to ten thousand feet per hour." The impatient cranker was an artistic hazard; also, said Phalke, a destroyer of prints. *Evidence,* III, 872–73.

had been interesting but remote. The slinky French heroine Protea, the Italian comedian Foolshead, and those American Keystone Kops were all amazing but might as well have come from Mars. In the Phalke films the figures of long-told stories took flesh and blood. The impact was overwhelming. When Rama appeared on the screen in *Lanka Dahan,* and when in *Krishna Janma* Lord Krishna himself at last appeared, men and women in the audience prostrated themselves before the screen.

In due time Phalke, like other producers of his period, became an exhibitor and traveled far and wide by bullock cart, with projector, screen, and films. The people who came were seldom two-rupee customers. Most paid four annas, two annas, or even one anna, and most of them sat on the ground. The revenue was in coins. The weight of the coins, on the homeward trip, could be enormous.

Rajah Harischandra ACTORS INCLUDING SALUNKE, MALE HEROINE

THIS AND FACING PICTURES OF PHALKE ARE FRAMES FROM FOOTAGE FOUND IN HIS HOME LONG AFTER HIS DEATH, DEPICTING WORK ON *Rajah Harischandra*—PROBABLY ON 1917 REVISIONS OF THE FILM. THE FOOTAGE MAY HAVE BEEN PART OF A LOST PHALKE DOCUMENTARY ON *How Films Are Made.*

PHALKE IN HIS STUDY

The success of the Phalke films extended eventually to all parts of India. The showing of *Rajah Harischandra* in Madras brought, according to one observer, "an almost phenomenal rush."[35] A showing of *Krishna Janma* in Madras was described by the exhibitor in these terms: "At one time . . . where the road is 100 feet broad in front of my theatre, the whole road was blocked with traffic."[36] Some of the Phalke films were shown to Indian audiences in Burma, Singapore, and East Africa.[37] In the case of *Lanka Dahan,* one exhibitor gave showings every hour from 7 A.M. until after midnight.[38] *Lanka Dahan* and *Krishna Janma* remained in circulation for over a decade.

Dadasaheb Phalke was of short stature and unassuming manner, but quick in movement. He was by all accounts an indefatigable worker: producer, director, writer, cameraman, scenic artist, make-up man, editor. In one short film, *Professor Kelpha's Magic,* he performed as magician.

Rajah Harischandra was filmed in Bombay and vicinity while the Phalkes lived on Dadar Main Road, now Dadasaheb Phalke Road.

[35] *Evidence,* III, 280. [36] *Ibid.,* I, 356. [37] *Ibid.,* II, 874.
[38] Suresh Phalke, "The Film Industry and Phalke," *Hindustan Standard,* February 24, 1961.

PHALKE BUILDING SETS FOR *Rajah Harischandra*

PHALKE EDITING *Rajah Harischandra*

After this first film Phalke moved his enterprise to Nasik, where all subsequent films were produced. Here the family lived in a three-story house, on a few acres of varied character. The family included Dadasaheb Phalke, Kaki Phalke, their five sons and three daughters, and relatives. All the children at one time or another appeared in Phalke films. Kaki Phalke, the mother, loaded and unloaded the camera, rushed film to the laboratory—a portion of the kitchen area—and supervised all laboratory work. As the company grew, she also supervised the cooking for the entire company.

Several dozen people worked on *Rajah Harischandra,* but during the following years the company grew to about a hundred.[39] All lived on the Nasik estate, the lesser members in various adjoining one-story buildings. The company became an extension of the joint-family system. All essential work was done by the members of this enlarged family. Except for an occasional crowd scene, no outsiders were involved. The Phalke enterprise thus set a pattern that was to dominate Indian film production for several decades.

The plot of land, which contained woods, hills, fields, and caves, provided a diversity of scenic backgrounds. Interior sets, open to the sunlight, were built in the garden behind the house. All "indoor" scenes were shot by sunlight; some use was made of reflectors.

Phalke was a stern disciplinarian, maintaining strict schedules and rules of procedure. Occasionally, infraction of rules brought instant dismissal. Phalke often discussed problems with members of the company, while reserving final decisions. At the Sunday midday meal, when the children were home from school, he would delight in putting problems to them. While planning *Lanka Dahan,* he explained that they would show a view of the sea from the Indian shore, supposedly looking toward the island of Lanka, that is, Ceylon. Then, later on, they would show the causeway built by the monkeys, stretching far into the sea in preparation for the invasion. He asked the children: How could they show that? In this way he would stimulate discussion on problems that were, presumably, revolving in his own mind. This particular problem was solved by Phalke by superimposure of a sea and a dike; in each of the superimposed shots, areas of the picture had been masked.[40]

[39] *Evidence,* III, 875. [40] Interview, Neelakanth Phalke.

MANDAKINI PHALKE AS THE BOY KRISHNA IN *Kaliya Mardan* (SLAYING OF THE SERPENT), 1919. HINDUSTAN FILM COMPANY

Fragments of a number of Phalke films have been preserved and in 1956 a reel of the fragments was assembled by the Indian Motion Picture Producers Association. The over-all structure of each film is forever lost, but the fragments show a fine pictorial sense and remarkable technical resourcefulness. Like another magician who became a film pioneer, Georges Méliès, Phalke was a special-effects genius. He explored a vast range of techniques, including animation. He experimented with color, via tinting and toning.[41] He used scenic models for a number of sequences, including the burning of Lanka, for which he also burned down two full-sized sets. He took interest in every detail of laboratory work. Having determined the right timing for the printing of a sequence, he set a metronome going to guide his wife: she turned the handle of the printer in time

[41] Tinting involved frame-by-frame brushwork on the print. Toning was done in the developing bath; through one or another chemical, an entire sea sequence was given a blue color, a fire sequence a red color.

with the metronome. He was often at odds with backers because he poured time, energy, and money into technical experiments. Sometimes these diverted him from film production, as when he developed a soap formula and launched a small soap factory, losing money on the venture.[42]

Not all his experiments were technical. He persuaded a Maharashtrian woman, Kamala, to play the leading role in *Bhasmasur Mohini,* his second feature production. His own daughter Mandakini played the boy Krishna in *Kaliya Mardan* (Slaying of the Serpent), produced in 1919. During the 1920s his company included a number of women. Thus, although fear of stigma remained for years a film industry problem in the casting of women's roles, Phalke took the first steps toward overcoming it.

In 1914 Phalke made a second trip to London, with his first three films. The proprietor of the *Bioscope* arranged a showing for a film industry group. The group must have been baffled by the content, so alien to its own preoccupations. But it apparently treated Phalke with considerable respect, and he was grateful for the attention he received.[43] The *Bioscope* expressed the opinion that "Mr. Phalke is directing his energies in the best and most profitable direction in specializing upon the presentation by film of Indian mythological dramas."[44]

Although resembling Méliès in technical skill, Phalke was never interested in amazement for its own sake. The material of his features came from a mythology that had, for its audience, religious meaning. To people unfamiliar with this material the films unquestionably seemed naïve. To those raised on the tales of Hindu gods and heroes, they opened a world of wonder. They earmarked for the Indian film an area of subject matter that won for it an immediate and powerful hold in India and neighboring countries—and at the same time shut it off from others.

The mythological film was to dominate Indian production for some years, but rival genres would begin to compete for attention. In the 1920s the film of modern background, the "social," rose in importance, and the Indian "historical" had a beginning. At the

[42] Interview, Neelakanth Phalke.
[43] *Evidence,* III, 879. [44] *Bioscope,* June 4, 1914.

CORONATION CINEMATOGRAPH
AND VARIETY HALL.
SANDHURST ROAD, GIRGAUM.
BEAU IDEAL PROGRAMME.

1½ Hours Show Throughout this week 1½ Hours Show

RAJA HARISCHANDRA.

A powerfully instructive subject from the Indian mythology. First film of Indian manufacture. Specially prepared at enormous cost. Original scenes from the sacred city of Benares. Sure to appeal to our Hindu patrons.

Miss IRENE DELMAR.
(Duette and Dance.)

THE McCLEMENTS.
(Comical Sketch)

ALEXANDROFF.
THE WONDERFUL FOOT-JUGGLER.

TIP-TOP COMICS.

Time:—6 to 7-30; 8 to 9-30; 10 to 11-30 and 11-45 to 1-15.

Note Double Rates of Admission

PHALKE RECORD
Bombay Chronicle, MAY 3, 1913: FIRST ADVERTISEMENT, FIRST INDIAN FEATURE FILM

FILM PRODUCTION IN INDIA.

Since one of the greatest and most valuable possibilities of the cinematograph is the circulation throughout the world of plays dealing with national life and characteristics, acted by native players amidst local scenes, it is with no small interest that one awaits the appearance in this country of Mr. D. G. Phâlke's first Indian films, some details of which were given in last week's BIOSCOPE. In spite of its wonderfully beautiful and distinctive qualities, Indian mythology is practically unknown in this country, with the exception of a few stories which have reached us through rather inadequate translations, and one feels, therefore, that Mr. Phâlke is directing his energies in the best and most profitable direction in specialising upon the presentation by film

of Indian mythological dramas. In a film version of a story, the whole beauty of the original may be retained so far as its action and characterisation are concerned, whilst it is possible to realise local colour and scenic detail in a manner which would be quite out of the question in any purely literary form, or even in the most lavish stage production. One feels, in short, that the cinematograph is the ideal medium for the presentation of all such stories, in which, if they are to be fully understood and sympathised with by foreigners, vivid realism of atmosphere and setting are essential considerations.

Mr. Phâlke's first film, "Harishchandra," was received with the greatest enthusiasm in India, the leading Bombay newspapers having reviewed it very favourably and at considerable length.

Bioscope, JUNE 4, 1914: VISIT TO LONDON
COURTESY NATIONAL FILM ARCHIVE, BRITISH FILM INSTITUTE

DADASAHEB PHALKE AND "KAKI" PHALKE

same time the "stunt film," inspired by the popular serials and by the features of Douglas Fairbanks, became an obsession with Indian producers. During this decade Phalke gradually began to feel like a stranger in the film world.

Rajah Harischandra, produced under the banner of Phalke's Films, was launched on a capital of Rs. 15,000.[45] Its success apparently made possible the subsequent productions of Phalke's Films, which included several features along with short documentaries, "topicals," trick films. The titles of several lost shorts of this period suggest Phalke's experimental bent: *Laxmicha Galicha* (Animated Coins), *Agkadyanchya Mouj* (Game of Match Sticks). Ahead of his time, he felt audiences should understand the film medium, and so he produced *Chitrapat Kase Tayar Kartat* (How Films Are Made), released in 1917. That year he also issued a revised version of *Rajah Harischandra.*

[45] *Evidence,* III, 878.

But rising costs, needs in new equipment, and the growth of Phalke's production ambitions all pressed him to search for new capital. About 1917 five new partners, including Mayashanker Bhatt, a textile manufacturer,[46] entered the picture and the company was reorganized as the Hindustan Film Company. Phalke soon quarreled with the new partners, and for two years, 1919–21, retired to Banaras.[47] Then he returned to Nasik, resolved the quarrel, and resumed work—though more often as production supervisor than as director. Film tastes were changing, and the competition seemed to call for rapid and steady production in an increasingly commercial atmosphere. In 1928 Phalke told a government committee: "I have retired, being disgusted with it—I do not care so much for money. I care more for technicality and first class production. I could not succeed so I left it off."[48]

In 1931 he tried again. Backed once more by Mayashanker Bhatt, he produced *Setu Bandhan* (Bridging the Sea). Coming at the last moment of the silent era, it was ill-timed. Phalke tried to salvage it by postsynchronizing dialogue, the first such effort in India. Then he made a talkie, *Gangavataran* (The Descent of Ganga), for another company. But the tide was no longer with him. Phalke lived until 1944. When he died at Nasik, on February 16 of that year, he was almost forgotten. During his final months all memory of his days of fame left him.

Phalke had laid the cornerstone of an industry. The Indian film world measures its existence from the release of *Rajah Harischandra,* India's first feature, in 1913. That film—the spell it cast, the crowds it drew—persuaded many a tent showman, many a cinematographer of topicals, many a backer, to take a fling at the feature film. Within a few years film production broke out like a rash in many parts of India.

[46] *Indian Cinematograph Year Book,* 1938.
[47] Now Varanasi.
[48] *Evidence,* III, 883.

Three Get Started

Dhiren Ganguly—sometimes known by a longer version of his name, Dhirendra Nath Gangopadhaya—was born in 1893 in Calcutta. He studied at the University of Calcutta, then went to nearby Santiniketan to pursue art studies under its already famous founder, Rabindranath Tagore, whose spirit found expression in poetry, story, drama, essay, music, dance, and painting. From these studies Ganguly went to a position in Hyderabad, in the heart of the Indian subcontinent, at an art college sponsored by the Nizam of Hyderabad, one of the most powerful of Indian princes. Heir of a long line of Nizams that went back two hundred years, this potentate was virtually all-powerful in the vast area under his control. A Muslim, he ruled over a population of 16 million, of whom about 13 million were Hindus. He was often referred to as the richest man in the world and the size of his harem was a subject of rumor and legend. Under the existing relationship with the British, the Nizam of Hyderabad, like the more than 500 other rulers of "princely states," recognized the "paramountcy" of the British, but this scarcely affected his internal authority. Often benevolently exercised, it was close to absolute.[1]

At the Nizam's art college Ganguly soon acquired the title of headmaster. But he had time for other projects. Interested in acting and photography, he published in 1915 a book of photographs, *Bhaber Abhibaktae* (Experiments in Expression), in which he himself appeared in a vast variety of roles: men and women of all ages and all segments of society. In some photographs he appeared in several guises. In one such photograph, for example, he was an orator on a soapbox and also each of the four people listening. The book provided an outlet for a rich satiric sense and was instantly popular, running to several editions and leading to additional vol-

[1] Wallbank, *A Short History*, pp. 16–17.

umes. *Bhaber Abhibaktae* was followed by *Amar Desh* (My Country) and two other sequels.[2]

Ganguly sent the first of these books to J. F. Madan in Calcutta, expressing interest in the new art of the motion picture. Madan immediately encouraged him to come to Calcutta for a talk. Madan by now owned all the theatres in Calcutta except one, the Russa, as well as theatres in a number of other cities. He was an importer of film equipment and films, which he distributed to his theatres and others. Along with film interests he also had a Calcutta theatrical company and was starting to produce films of various lengths based on its productions. He was also making occasional topicals. When Ganguly came to see him about 1918, Madan was interested in Ganguly's acquaintance with Tagore, and encouraged him to get the poet's permission to make a film based on Tagore's play *Sacrifice*. Ganguly went to Tagore, who promptly gave his consent.

But the *Sacrifice* film was postponed by other developments. A Calcutta businessman, P. B. Dutt, who had made substantial profits from the manufacture of wooden buckets, wanted to invest in the film field. He suggested to N. C. Laharrie, of Madan's organization, that he leave Madan to form a new unit. Ganguly became a member of this group. Named the Indo-British Film Company, it consisted of four partners: Dutt as financier, Laharrie as general manager, J. C. Sircar as cameraman, and Ganguly as "dramatic director," which apparently included writing. Ganguly promptly wrote a story for them and by 1920 they were shooting it, with Ganguly playing the leading role. The film opened the following year in the one film theatre in Calcutta not owned by Madan, the Russa, where it was a resounding success. A zestful comedy, *England Returned,* it satirized the pretensions of Indians back from England, full of Western ideas; at the same time it satirized the conservatism of those Indians to whom all new ideas were unwelcome. Impartial in its laughter, it escaped the stigma of propaganda. Produced

[2] Aside from launching a film career, these volumes led later to employment of Ganguly by the Calcutta Police Department to train detectives in the art of disguise. Decades later he was recalled to give similar instruction to the police of independent India. Interview, Ganguly.

PHOTOS THAT LAUNCHED A CAREER

FROM GANGULY'S BOOK *Bhaber Abhibaktae* (EXPERIMENTS IN EXPRESSION), 1915

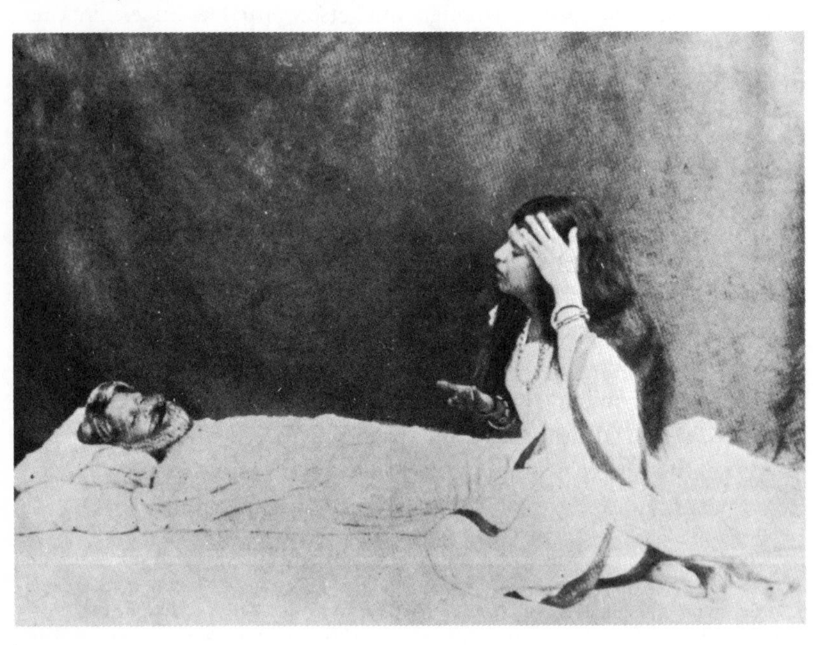

FROM GANGULY'S SEQUEL, *Amar Desh* (MY COUNTRY), CA. 1920
ALL CHARACTERS BY GANGULY

at a cost of Rs. 20,000, it earned more than this in its three-month run at the Russa. The Bombay rights were then sold to a Bombay theatre group for Rs. 22,000. J. F. Madan, ever the businessman, bought all remaining rights.[3] These profits put the partners into an exuberant state, so much so that within a year they parted company and went separate ways. Ganguly, by this time married to a distant relative of Rabindranath Tagore, returned to Hyderabad to head a new venture of his own, taking several Calcutta film technicians with him. The result was the Lotus Film Company, which began in 1922 under the benevolent eye of the Nizam.

The company set up its own laboratory and within a short time was also operating two Hyderabad cinemas. For the productions, the Nizam gave permission to use palace backgrounds.[4] The com-

[3] *Evidence,* II, 640–41. [4] *Ibid.,* II, 640.

pany got a rapid start and produced a number of films in quick succession. Some, like *England Returned,* were comedies and had English titles: *The Lady Teacher* and *The Marriage Tonic.* There was also a mythological, *Hara Gouri.* Another film, *The Stepmother,* was based on a Bengali play. Things were starting well for the Lotus Film Company.

In 1924 it offered, at one of its Hyderabad theatres, a Bombay-produced film called *Razia Begum,* based on historic events and telling of a Muslim queen who fell in love with a Hindu subject. The mid-1920s were a time of rising Hindu-Muslim tension. The makers of *Razia Begum* may have thought of their film as a contribution to interfaith amity. It had run successfully in Bombay, but immediately after its appearance in Hyderabad, a functionary of the Nizam arrived at the Lotus Film Company door. Ganguly and associates were instructed to leave the Nizam's domain within twenty-four hours.[5]

That day two theatres were closed, equipment was packed, and families and technicians departed. After stopping briefly in Bombay, Ganguly made his way back to Calcutta. Shipwrecked for the moment, he began, after a time, to try to organize a new venture. We shall leave him now but shall meet him again presently. He had already injected a new note of comedy into the Indian film, and burned his fingers on history. He would not be the last.

A stall in a bazaar

Debaki Kumar Bose, son of an attorney, was born in Akalpoush, in the Burdwan district of West Bengal—not far from Calcutta—in 1898. In 1920 he was busy with college studies and would soon take the University of Calcutta examinations that would make him a Bachelor of Arts. But 1920 was also the year that the Indian National Congress met in special session in Calcutta.

The most important figure at the meeting was Gandhi. Already a revered leader, he was now emerging as a great unifying force in the independence movement. Throughout the First World War he had cooperated with the British and urged faith in British assur-

[5] Interview, Ganguly.

With Compliments of
RUSSA THEATRE.

India's Greatest Expressionist
MR. DHIRENDRA NATH GANGOPADHAYA.
AS THE
"ENGLAND RETURNED"

HANDBILL FOR *England Returned,* 1921. INDO-BRITISH FILM COMPANY

ances of postwar reforms and progress toward self-rule. It was, after all, through British rule that democratic aspirations had taken root in India.[6] When in 1918 a reform plan was proposed in the Montagu-Chelmsford Report, Gandhi had at first seemed ready to consider it. But in the end he condemned the program as inadequate, a mockery of Indian hopes. The proposals, nonetheless, became law in December, 1919.

When the Congress convened in Calcutta the following year, other problems were contributing to the growing tension. The year 1919 had seen a continuation of rising prices, bringing destitution and hunger to millions. The year had also seen enactment of the Rowlatt Act, which gave the Government of India, in emergencies, the right to judge cases without trial. Defended as an essential shield against bolshevism and anarchism, this measure seemed to many a move to throttle the independence movement. Gandhi's protests resulted, to his dismay, in widespread riots and bloodshed, which led

[6] Kabir, *Britain and India,* pp. 3–13.

to the brief imprisonment of Gandhi, still bloodier riots, and still more drastic repressions, culminating in the Amritsar massacre. When a crowd had gathered in Amritsar, in the Punjab, in spite of a proclaimed ban on gatherings, General Dyer had ordered his troops to fire without further warning; four hundred people had been killed and more than a thousand wounded. Protests were worldwide. It was with the vivid memory of such events that the Congress met and listened to new proposals by Gandhi.

Gandhi proposed, and the Congress soon afterwards adopted, a program of nonviolent noncooperation that included (1) surrender of British titles; (2) refusal to attend government functions; (3) withdrawal of students from schools and colleges; (4) boycott of courts; (5) refusal to serve in British forces in the Near East; (6) noncooperation with announced government reforms.

These were the reasons why young Debaki Bose in 1921 did not take the examinations for which he had long prepared but instead walked out of his college. His father, an attorney who had maintained good relations with British officialdom, was outraged by this behavior and cut him off. Debaki Bose went to Burdwan and opened a stall in a bazaar, selling napkins. His own relatives avoided the sight of this degradation. Meanwhile he also became assistant editor of a Congress weekly, *Sakti* (Strength). For some years he made a living by whatever means he could. Then, in the mid-1920s, Dhiren Ganguly came to Burdwan to try to sell shares in a new film company. A physician who invested urged Ganguly to interest himself in Debaki Bose. Ganguly suggested that Bose write a script, and Bose soon thereafter sent him the script for *Flames of Flesh,* which eventually became the first production of the British Dominion Film Company.

Debaki Bose went to work for British Dominion for Rs. 30 a month. Besides writing its first feature, he played the leading role. For this ambitious production several dozen people traveled to Jaipur, where the Maharajah of Jaipur not only let them use the famous Amber Palace as background but loaned horses and elephants. When the film opened in Calcutta, Debaki Bose sat behind the screen, directing a group who made sound effects of crowds and horses' hoofs. It was the beginning of a film career that was to continue for decades, and which would bring into the Indian film a special Indian note of dedication and fervor. Debaki Bose was a devotee, a Vaishnavite, who could speak freely about the film medium and what it could do in the cause of love, in a way that film makers of other nations would not be likely to do. Love, said Debaki Bose, begets love. Only love, he said, can "bring about fruition in all human efforts, including the making of films."[7] The sound film, especially through its resources of music, was to give him the opportunity to emerge as one of India's most notable directors, although in the end he despaired of the drift of his industry.

A film for Idd

Chandulal J. Shah was born in 1898 in Jamnagar, near Bombay. He studied at Sydenham College in Bombay and prepared for a career in business. After graduation, while looking for a position, he worked with a brother, D. J. Shah, who had written mythological films for several rising Bombay producers.

It was a time of tension and hunger but also of enterprise. The First World War had stimulated Indian business and industry. Before the war British policy had generally discouraged Indian industrialization; the function of India, in the colonial plan, had been to serve as a source of raw materials and a market for British manufactured goods. Although an Indian steel industry had been launched in the prewar years by the Tata family, its existence was regarded as precarious because of the possible hostility of Sheffield interests. But the First World War brought a change. The strain on British manufacturing made it desperately important to Great Britain that Indian industrialization be speeded. An Indian Munitions Board was set up in 1917 to make India, in large measure, "the arsenal for the Allies in the Near East."[8] India became an expanding source of steel rails, clothing, boots, tents, jute goods. The Board furthered expansion of wolfram mines, iron and steel works, cotton and jute mills. All this brought economic expansion to various areas, and

[7] Bose, "Films Must Mirror Life," in *Indian Talkie, 1931–56,* p. 43.
[8] Wallbank, *A Short History,* p. 124.

especially to the huge port of Bombay. Along with the officially sponsored growth there was other expansion. Wartime shortages brought speculation and black market trading, and these too put money into circulation in the big cities. All this had an impact on the infant film field, as similar conditions in the Second World War were to have years later, in even larger measure.

Funds came into film from a variety of sources. Hindustan Films, as we have seen, was made possible by a textile manufacturer, along with others. Jagadish, another new company of this period, was financed by a cotton merchant. In Calcutta an Eastern Film Syndicate was launched with the aid of a hair oil manufacturer.[9] In Bombay similar investments were creating other companies. In the words of one producer, the successes of Phalke "gave impetus to many capitalists in Bombay to rush to this industry."[10] Among the Bombay investors were the owners of theatres, who were by now competing vigorously for new films, especially the better Indian films. This trend is illustrated by the career of Chandulal J. Shah.

In 1924 Shah got a job on the Bombay Stock Exchange and felt he had settled down to a life of business. But the following year he heard that the Imperial Theatre was desperate for a film to be launched the week of Idd. For this Indian holiday, somewhat more than a month away, the theatre wanted an Indian film and had so far failed to obtain one. Chandulal Shah, silently aided by the reputation of his brother and his own vague association with several of his brother's mythological films, offered to have a film ready before the deadline. The theatre agreed to advance Rs. 10,000, half the usual budget for a 6,000-foot Bombay feature of the time. Within a day or two photography was begun, with Shah directing. When the theatre made enquiries two weeks later, he gave assurance that the film would be ready for the holiday. It was now that the theatre manager learned, to his horror, that Shah was not producing a mythological but a story of modern background. "I need a mythological!" he pleaded. "Something that will run at least a month!" A modern Indian story, he was sure, could not last more than two weeks. But it was too late to start over. Shah delivered the film be-

[9] *Evidence,* II, 691. [10] *Ibid.,* III, 678.

fore the deadline and, according to his recollection, it ran ten weeks.[11]

The following year, a similar crisis made Chandulal Shah irrevocably a film producer. He was watching a matinee at the Bombay Opera House when he was called from his seat. The Kohinoor Film Company, which had been producing since 1918, was in the midst of another production destined for the Imperial Theatre, which had advanced the principal capital. The director, Homi Master, had fractured his ankle and in his agony kept asking for Shah. That afternoon Shah visited Master at the hospital. The latter took a script from under his pillow. It was, he said, the climax of his career, his life ambition. "You will have to finish it." Shah protested that Master must finish it himself, later. But Master said that a film starring the beautiful Miss Gohar, who had been a sensation in another Kohinoor production, had been promised to the Imperial and was due in twenty-one days. It was half finished; there was no turning back.

Shah had no interest in Master's script, a love story based on Indian mythology, but he was desperately interested in Miss Gohar, whose "expressive eyes" he had already admired. He undertook to fulfill the contract with the Imperial, but wrote a new scenario that night. By the following day he had scheduled photography with Gohar. During the following weeks he shot every day and, as footage became available, edited at night. He finished, in international film tradition, with a fever of 104 degrees, before the deadline date. The film, *Gun Sundari* (Why Husbands Go Astray),[12] was a milestone in the rise of the Indian social film. It was later remade by Shah as a talkie in three different Indian languages and each time was a box-office success. It also began for him an association with Gohar which was to last throughout a long professional career.

The role of the social film in India, and Shah's contribution to it, are readily indicated by a synopsis of *Gun Sundari*. In a Western set-

[11] Interview, Shah.

[12] English titles in parentheses are in most cases translations of the Indian titles. But in some cases, as in this instance, the title is an official English-language title —not necessarily a translation—used by the producer for English-language newspapers and other promotion. During the silent period many Indian films had such alternative titles.

"GLORIOUS GOHAR" in *Devoted Wife,* 1932. REMAKE, WITH SOUND, OF 1925 *Gun Sundari* (WHY HUSBANDS GO ASTRAY). RANJIT

ting its plot would be conventional; in an Eastern setting it was a sensation—exciting to some, deeply disturbing to others. It might have been suggested by any of the innumerable American films coming to Bombay's Empire, Excelsior, Capitol, Universal, and Royal theatres. In Indian dress it was a challenge to Indian ways.

It concerned the dilemma of the dutiful wife. She is shown to be involved in all the problems of the household, of all the members of the family. At night she takes those problems to the bedroom. But her husband, who has his own work problems, doesn't want more problems. That is why he turns to the dancing girl. The film said, in effect: "Don't be only a dutiful wife. Be a companion too." Early in the film there was a scene in which the husband prepares to go out in the evening and the wife asks where he is going. He replies that nowadays a wife doesn't ask her husband such a question. During the following weeks a change comes over her, she becomes a modern woman. One evening, dressed to go out, she is asked by her husband where she is going. She replies that nowadays a husband doesn't ask his wife such a question. As a result of all this she becomes her husband's companion, not "only" his dutiful wife.

Gun Sundari was followed in the same year by *Typist Girl.* Like several other Chandulal Shah social films, it had *only* an English title. This was logical, because again the drama lay in the transfer to an Indian world of elements of Western life—more accurately, of the Western film. *Typist Girl* again featured "glorious Gohar" and also an Anglo-Indian girl, Ruby Meyers, who left her job with the tele-

Miss 1933, WITH GOHAR AND E. BILLIMORIA, 1933. RANJIT

GOHAR IN *Rajputani* (RAJPUTANI), 1934. RANJIT

phone company to become the actress Sulochana, for two decades an idol of movie-going India and still active in the 1960s.

Gun Sundari and *Typist Girl* put the Indian social on a footing with the mythological—at least, as far as the urban Indian was concerned. Later Chandulal Shah was to have his own company, Ranjit Films, build studios and a large organization, and produce some 130 features, including more than sixty money-making films in a row—all of them socials.

It has often been a source of exasperation to Indian producers that films of the sort produced by Chandulal Shah, of sensational impact in an Indian environment, were viewed by Westerners as mild and hackneyed. Between two societies so far apart, the problem film is not readily exchangeable. In time Chandulal Shah was to travel to Hollywood, win admiration for his charm and his record of successes, have his picture taken with starlets, and discuss co-production deals with American executives—but a meeting of minds would be difficult.

"Had a short life"

We have traced the migrations of three men through the early film world of India. Their work was done against a background of turbulent times. The rising nationalism of India was building pressure. With the sense of growing power, of the inevitability of victory, old internal tensions were also rising to the surface. As we have seen, Hindu-Muslim relations were deteriorating in the 1920s. If little of the headline issues of the age appears in the film work of the three men we have discussed, there are ample reasons for this—as we shall see later.

These men were three among thousands. They survived the competition while others did not. Without sufficient funds, equipment, or talent, many fell by the wayside. In 1933 a film enthusiast published the first of various Indian film annuals and "who's who" volumes that have appeared from time to time, and faithfully listed countless companies of previous years, only a few of which still existed. They had sprung up in Bombay, Calcutta, Madras, Kolhapur, Hyderabad, Lucknow, Gaya, Delhi, Ahmedabad, Peshawar,

Secunderabad, Nagercoil. In listing vanished companies, the editor sometimes added cryptic explanations or comments. For Bombay he listed Oriental Pictures Corporation ("Had a short life"), Young India Film Company ("One picture and then died"), Jagadish Films ("Defunct"), Excelsior Company ("Shut down"), Suresh Film Company ("Liquidated").

In Calcutta he listed the Indo-British Film Company ("Broke up"), Taj Mahal Film Company ("Short-lived"), Photo-Play Syndicate of India ("Flashed like a lightning and as quickly disappeared after their first picture, *Soul of a Slave*"), Eastern Film Syndicate ("Low moral tone stood in the way of their second picture, *Bicharack,* released after cuts—collapsed"), British Dominion Films ("Collapsed due to internal troubles"), Heera Film Company ("Not now functioning").

As to Madras, the list told of Nataraja Mudaliar ("Made a bold stand about a year or so producing mythological films. . . . His pictures were bad from all standpoints"), Star of the East Films ("Wound up owing to lack of capital"), Guarantee Film Company ("Guaranteed pictures no doubt but strange she did not guarantee her life"), General Pictures Corporation Ltd. ("Liquidated"), Associated Films Ltd. ("Failed for want of business-like instinct").[13]

Many of these enterprises had started with only the sketchiest technical preparation. A few had started on the basis of correspondence courses given by one or another "institute" in the United States.[14] Some started on the basis of one man's travel and observation abroad. In 1921 one young Indian, in London, sought permission to watch production at one of the studios and had been asked to pay a "premium" of £1,000, which he could not afford. He went to Germany and secured the same privilege for a more modest £15 per month. "The only training I got there was I saw how some of the well-known experts were directing and how things were carried out."[15] One man had traveled to the United States in the hope of making such observation but gained entry, after long effort, only by becoming an extra. And on the basis of another man's camera experience in the United States, a Bombay company was formed and

[13] *Who Is Who in Indian Filmland,* pp. 4–11.
[14] *Evidence,* III, 327, 364. [15] *Ibid.,* I, 181.

began production, "when unfortunately he died and the company had to go into liquidation."[16]

Nevertheless, enterprises were launched. It is not surprising that various observers were saying, during the 1920s, that Indian films were becoming worse, not better. It is also not surprising that by the end of the 1920s, capital was becoming scarce.

Among exhibitors, too, there was mushroom growth and high mortality. The number of theatres in India increased from about 150 in 1923 to about 265 in 1927.[17] This brought a sharply increased demand for Indian films, but the supply of usable films could not meet it. As for foreign films, an obstacle was Madan Theatres Ltd. By 1927 its chain numbered 85 theatres–65 owned and 20 supplied under contract.[18] Such a chain of theatres could and did outbid all other exhibitors for the best foreign films. Exhibitors were often faced with nightmare uncertainties about film supply, and sometimes took foreign films they did not want.

This reminds us that the Indian films of the 1920s were only a part of what the Indian filmgoer was seeing. In the other part, the foreign supply, an important change was taking place.

[16] *Ibid.*, I, 13.
[17] *Report of the Indian Cinematograph Committee,* p. 179.
[18] *Evidence,* II, 828. These figures include Madan theatres in Burma and Ceylon.

Empire

We have noted that when Phalke began his work, in the years before the First World War, Indian cinemas were showing an international assortment of films. This was true also of theatres in Great Britain, the United States, and other countries. In 1910 the features released in Great Britain included 36 from France, 28 from the United States, 17 from Italy, 15 from Great Britain, 4 from Denmark, Germany, and elsewhere.[1] The films shipped to India before the war, and during its first year or so, reflected this pattern. Then the change came.

The outbreak of war in 1914 almost stopped film production in France and Italy, handcuffed English production with scarcities and restrictions, and isolated the German studios. But audiences everywhere remained ravenous for films, which were suddenly regarded as necessary for morale. American producers, now establishing themselves in Hollywood, were ready to fill the need. A fantastic American expansion began, which soon made Charlie Chaplin, Mary Pickford, and other emerging stars household deities throughout the world, created fortunes, and set the stage for further expansion after the war. By the time the treaty of Versailles was signed Hollywood was the world film capital. Trench warfare was over. But the international film struggles were only beginning.

For the other film-producing nations, the problem was not only one of physical recovery. During the war the United States had set the pattern of film distribution. As early as 1915, Great Britain had portents of what this might mean. The Essanay company, controlling the most-wanted of all films, those of Charlie Chaplin, began to require British exhibitors to take the whole Essanay output along with Chaplin. British producers found British theatres booked far ahead by this "block booking" and increasingly unable—or unwill-

[1] Low, *History of the British Film, 1906–1914,* p. 54.

WEDNESDAY, FEBRUARY 25, 1920. 5

ENGAGEMENTS.

THE EXCELSIOR.

TO-DAY at 6-30 p.m.

"SPLENDID COWARD"

A Comedy Drama in 6 Reels.

The Story of a Son's sacrifice to shield his Mother.

THE EMPIRE.

DAILY at 6-30, 8-15 and 10 p. m.

CONCLUDING CHAPTERS OF

"INTOLERANCE."

THE THEME

"Why touch upon such themes!" perhaps some friend
May ask, incredulous; "and to what good end?
The errors of an age long passed away?
I answer: "For the lessons that they teach;
The tolerance of opinion and of speech."

—*Longfellow.*

LOVE'S STRUGGLE THROUGHOUT THE AGES

ENGAGEMENTS.

Messrs. K. D. and Bros. present to their patrons, to-night, a film of GLORY, POMP and GRANDEUR!

These are the Three keynotes of the THEDA BARA Super-Picture.

"CLEOPATRA"

that has created and is daily creating—a furore in the City.

One's hair seem to rise as you watch past History unfold itself in a great, breathless panorama. Rome, Alexandria and Egypt of the Barbaric Ages live again. You see the magnificent great scenes of the triumphal entry of Caesar to the Forum,—the oration over his dead body by Mark Antony.—several thousand men engaged in the bloody Battle of Actium.—the Ancient Ships of War ablaze with battle fire,—the encamping on the desert of the huge army of the Siren,—the Pyramids, the Sphinx, the tremendous city strewn over with thousands of tents! All this holds you wonderbound by their artistic, marvellous, gorgeous, graphical accuracy and detail!

The WILLIAM FOX VERSION of

"CLEOPATRA"

is a picture in a Million.

And it has cost a Million Dollars—3,000,000 of Rupees!—Fancy that! The

ENGAGEMENTS.

TWO UNIVERSAL SUPER-SERIALS, await you this week at the palatial Picture-House,

CINEMA PRECIOUS,

(Grant Road Tram Terminus).

If you are fond of Thrills, Mystery, Romance, Adventure, begin right in earnest with our monster Double Programme that has commenced only last Saturday and is going through its very first instalment.

"THE MIDNIGHT MAN"

(5,000 feet—Opening Chapters) and the

"SPURS AND SADDLE"

(4,000 feet—Opening Chapters.)

starring the best known, the best loved Athletic Hero in all the world, JAMES CORBETT, known everywhere as the "Gentleman Jim"—the refined, popular big brother of boys from 8 to 80. And featuring the darling of the Cinema Circles, the infatuating MARIE WALCAMP—the Cinema Idol in Europe, America, Australia, Asia and Africa. Thunderous applause and continuous hurraying greet this favourite couple, Show after Show, at the

CINEMA PRECIOUS,

(Grant Road Tram Terminus).

These Two Films are supplied by the Universal Film Manufacturing Company.

Four Shows To-night: 5-30, 7, 8-45, 10-30 p.m (S.T.)

CINEMA 1920: U.S. DOMINATION
Times of India, COURTESY ASIATIC SOCIETY OF BOMBAY

ing—to absorb the slim output of the British studios. A plan to require theatres to show a minimum quota of British films was first proposed in 1917 but failed of adoption. By the end of the war, in 1918, the British film industry "found the American stranglehold too strong to break."[2] By 1925 it was estimated that 95 percent of British screen time was occupied by American films.[3] In France, a similar situation prevailed.

If British, French, and other producers were finding it difficult to regain a toehold on their home grounds, they now found their former United States markets even more impenetrable. Here vast consolidations were taking place. In some cases theatre chains, such as Loew's, were purchasing studios in order to be certain of a steady flow of films. In other cases producers, like Paramount, launched theatre-buying and theatre-building programs in order to have a secure home market. Paramount, starting such a program in 1919,

[2] Balcon *et al., Twenty Years of British Film,* p. 13.
[3] Low, *History of the British Film, 1914–1918,* pp. 65–66.

had 300 theatres by 1921 and almost a thousand a decade later.[4] Fox and Warner Brothers also bought American theatres by the hundreds; Universal followed the example on a more modest scale. Many theatres not purchased came under the control of the producers via block-booking contracts. Opportunities for foreign films became severely restricted. But the large American producers were now secure in their home base, and amply supported by it. Foreign markets came to represent pure profit.

In India this meant that American films could always be offered at lower prices than most other films, including Indian films. An Indian film usually had to recoup Rs. 20,000 in its home market. The importer of an American film could usually purchase Indian distribution rights for a fraction of this. In 1927 an importer of some Columbia Pictures productions paid as little as Rs. 2,000 per feature for rights in India, Burma, and Ceylon,[5] although most prices were higher.

In 1916 Universal became the first of the American producing-distributing companies to establish an agency in India. By the mid-1920s it was offering Indian theatres 52 features, 52 comedies, 52 newsreels per year. Block booking seems to have been involved in some cases but not in others. Universal appears to have felt that the Indian market was worth nursing patiently, and it won among some exhibitors a reputation for humanity. One exhibitor, irate at the demands of film distributors, declared: "The noblest exception to this statement is the Universal Pictures Corporation, whose agent in Bombay and the several local managers are very considerate to the theatre owners."[6]

Pathé-India had been established in Bombay as early as 1907 as the concessionaire for the films of Pathé Frères.[7] Alex Hague, for over two decades manager and sole proprietor of Pathé-India, could also handle other products and became the Indian distributor of First National Pictures—later absorbed by Warner Brothers. Pathé-India became an importer of American more than of European features.

During the 1920s the products of some of the other American

[4] Conant, *Antitrust in the Motion Picture Industry*, pp. 25–26.
[5] *Evidence*, III, 435. [6] *Ibid.*, III, 384. [7] *Ibid.*, I, 503.

companies were imported by Madan Theatres Ltd. It appears to have purchased American films in wholesale lots to secure the outstanding big-name attractions. American films formed the staple of most Madan theatres. In 1923, on the death of J. F. Madan, control of Madan Theatres Ltd. passed to his five sons—B. J., F. J., J. J., P. J., and R. J. Madan. At that time 90 percent of Madan imports came from the United States, the remaining 10 percent from Great Britain, France, and Germany.[8]

In 1926–27 15 percent of the features released in India were Indian, 85 percent were foreign. Most of these were American.[9] The position of the foreign film in India was of course irksome to Indian producers. The dominance of American films among these imports was especially nettling to British producers.

In the postwar years Germany was the first country to strengthen the international position of its film industry through government action. Partly a continuation of its wartime mobilization of film, this action involved lavish government investments in studios and equipment, as well as production subsidies.[10] In the 1920s the German film underwent a dramatic rebirth, which had its impact in India in several German-Indian co-productions, starting in 1925 with *The Light of Asia,* which we shall discuss later.

In 1927 Great Britain at last moved to bolster its film industry. Its Cinematograph Films Act of 1927 was described as "an act to restrict blind booking and advance booking of cinematograph films, and to secure the renting and exhibition of a certain proportion of British films, and for purposes connected therewith." For British theatres the quota was to start at 5 percent and in a few years rise to 20 percent. Its purpose was achieved with remarkable speed. In 1926 Great Britain had produced only 26 feature films. Production rose to 128 in 1929, and to 153 in 1932.[11]

While making sure of a share of its home market, Great Britain was also thinking about its place in the film market of British India. On October 6, 1927, the Government of India announced the ap-

[8] *Ibid.*, II, 863.
[9] *Report of the Indian Cinematograph Committee,* p. 188.
[10] Kracauer, *From Caligari to Hitler,* pp. 35–39.
[11] Balcon *et al., Twenty Years of British Films,* pp. 14–15.

pointment of a committee of enquiry, the Indian Cinematograph Committee.

"Much harm was being done"

The background of this action was summarized by the committee in these words:

> Letters and articles have appeared from time to time in the British Press asseverating that much harm was being done in India by the widespread exhibition of Western films. We have seen several of these Press comments from 1923 onwards. The general trend of them is that, owing to difference of customs and outlook, Western films are misunderstood and tend to discredit Western civilization in the eyes of the masses in India. Such criticism was chiefly directed against "cheap American films." To give an example of this sort of criticism, a well-known Bishop intimately acquainted with India stated (as reported in the Press) in a speech at a conference in England in 1925: "The majority of the films, which are chiefly from America, are of sensational and daring murders, crimes, and divorces, and, on the whole, degrade the white women in the eyes of the Indians."[12]

In view of all this, the Indian Cinematograph Committee was instructed to study the adequacy of censorship as practiced in India and the need for stricter measures. And the committee had a further task:

> At the same time the question has been raised by a resolution of the Imperial Conference of 1926 whether the various parts of the Empire could take any steps to encourage the exhibition of Empire films. As all Governments of the Empire have been invited to consider this question, it appeared to the Government of India that it would be appropriate that it should be examined by the proposed Committee. This extension of the scope of the Committee's enquiry would also enable it to address itself to a question which may have a far-reaching influence on the development of the cinematograph in India, namely, the possibility of encouraging the production and exhibition of Indian films.[13]

Great Britain's careful approach to this problem, and the delicate wording of the resolution, reflected the nature of the relationship that existed in 1927 between Great Britain and British India.

[12] *Report of the Indian Cinematograph Committee,* p. 3. [13] *Ibid.,* p. xii.

Dyarchy in action

The reforms of 1921 had been put into effect in India despite the boycott of the Indian National Congress, and a certain amount of democratic machinery was now operating. In the provinces some departments had been "transferred" to Indian authority, and in various provinces British officials were working under Indian ministers. Meanwhile the franchise had been extended, and in one of the branches of the central legislature the majority of members were now chosen by the broadened electorate.

However, these democratic mechanisms were combined with others that left ultimate authority firmly in British hands. While some provincial departments had been "transferred," more crucial matters such as justice, police, and prisons were still "reserved." And both in the provinces and in the central government, the executive had not only absolute veto power over legislation but also independent legislative power. If the legislature failed to enact a law he considered necessary, the Governor-General—or, in the provinces, the British-appointed governor—could "certify" it as essential and so make it law.[14] These devices were the crux of Indian opposition to the reform plan. In the view of many Indian leaders this system of "dyarchy," as it was called, invited Indians to share responsibility without authority. Yet the machinery had been launched, and the British were at pains to emphasize its democratic aspects.

The Indian Cinematograph Committee was entirely in the spirit of the times. It consisted of three British and three Indian members. One of the Indians, Diwan Bahadur T. Rangachariar, a prominent Madras lawyer, was the committee chairman. The arrangement gave Indian members a preponderant dignity without a majority. The committee would, of course, make no ultimate decisions; it would study and report. Through its enquiry, it was asked to lay the foundation for protection of "Empire films."

For British purposes the resolution was well worded. The phrase "Empire films" was elusive, but the committee was urged to consider it as including Indian as well as British films. There was a spirit of partnership about this.

[14] Wallbank, *A Short History*, pp. 148–52.

In stating the problem in terms of a Western threat to Indian ways, the resolution was of course echoing a favorite theme of the Congress, and especially of Gandhi himself. In a characteristic utterance Gandhi had declared: "India's salvation consists in unlearning what she has learned during the last fifty years. The railways, telegraphs, hospitals, lawyers, doctors and such-like have all to go."[15] The committee was now asked to consider whether "such-like" did not include "Western films . . . chiefly from America."

To arrive at its decision the committee set out to study all aspects of film production, distribution, and exhibition in India, public reaction to them, and the operation of governmental supervision. It thus launched a major investigation, in the course of which it held hearings in a dozen cities, traveled 9,400 miles, visited production companies and theatres, questioned 353 witnesses, studied 320 replies to the 4,325 questionnaires it had issued, and spent Rs. 193,900.

The committee thought it worth recording that its witnesses had included 114 Europeans, Anglo-Indians, and Americans, and 239 Indians. Of the Indians, 157 were Hindus and 82 non-Hindus; the latter included 38 Muslims, 25 Parsis, 16 Burmese, 2 Sikhs, and 1 Christian. The committee also noted that it had examined 35 ladies, of whom 16 were Europeans and 19 Indians. The witnesses included members of the film industry and nonindustry people.[16]

In May, 1928, the committee completed its report, which was followed by a Minute of Dissent by its three British members, and the chairman's reply to the dissent. The report was printed, as was a transcript of all open hearings. The resulting material—the one-volume *Report of the Indian Cinematograph Committee* and four volumes of *Evidence*—forms a rich storehouse of information on the early Indian film.

Among the film-industry witnesses were Dadasaheb Phalke, Dhiren Ganguly, J. J. Madan, Alex Hague, Sulochana or Ruby Meyers, and other early film leaders not yet mentioned such as Himansu Rai, producer of *The Light of Asia*. Other witnesses included representatives of American companies, censorship officials, and Indian

[15] Quoted in Nehru, *Toward Freedom*, p. 314.
[16] *Report of the Indian Cinematograph Committee*, pp. 13–14.

exhibitors, including traveling exhibitors. The transcript provides many vivid word pictures of the Indian film world in action in its second decade.

In cities and mofussil

The theatres, as mirrored in the testimony, ranged from those of the Madan chain, one of which was about to install a Wurlitzer pipe organ at a cost of Rs. 65,000, to primitive cinemas in mofussil —the rural areas. Most theatres apparently had two or more showings a day; one theatre gave twelve a day during melas.[17] We learn that prices were usually in three or more classes, often from 2 or 3 annas to 2 rupees. In cities the top price might be 3 rupees, for "box" or "sofa" seats. In the lesser cinemas the lowest price might be 1 anna, for "ground" seats. In an Assam theatre the 393 tickets sold for one performance were for 350 ground seats, 40 bench seats, 3 chair seats; this was a normal distribution.

The films often had subtitles in three or four languages. A print made for circulation in the north might have each subtitle in Hindi, Gujarati, and Urdu; in the south a print might have subtitles in Tamil, Telugu, and English. Witnesses tell us that at each subtitle a rumble swept over the theatre, as people who could read proclaimed the words for those who could not. A few theatres had official readers.

A. We have translators now.

Q. Have you? On the cinema?

A. Oh yes. There is a man always standing there and explaining the film. He is a very clever fellow. He knows all about the story. Then as soon as one scene is on, he explains the whole thing in Telugu because everybody can't read what is on the film. He stands there throughout; he is a lecturer.

Q. We were told that such a man is a nuisance.

A. Not at all. He is paid 50 rupees.[18]

In a Northwest Frontier Province theatre this person was called a "demonstrator."[19]

[17] *Evidence,* I, 110. [18] *Ibid.,* III, 251.

[19] *Ibid.,* IV, 102. Never idolized like the Japanese *benshi* described in Anderson and Richie, *The Japanese Film,* pp. 22–34, the Indian narrators largely disappeared with the coming of sound.

Some of the mofussil theatres were described by witnesses as being in a sorry state:

> The lowest class of spectator has to squat on the ground and the benches and chairs in the other classes are in wretched condition and infested by bugs. There is no proper ventilation and most of the theatres are merely corrugated tin sheds. There is very little open space surrounding the theatre and no garden to please the eye and to attract the public.[20]

An exhibitor in Nagpur said his theatre had cost him Rs. 24,000. "The tin shed alone cost us 14,000 rupees. It is the biggest cinema in Nagpur."[21]

The city of Bombay had 20 cinemas, Calcutta 13, Madras 9, Delhi 6, Poona 6. A number of other cities had three or four.[22] Exhibitors testified to many problems with local authority: "The police, the custom, the postal, telegraph, municipal and a host of other people have to be admitted free to avoid trouble."[23]

Women filmgoers were scarce in the south and in mainly Muslim areas of the north but were increasingly evident in most cities. In Hindu areas mythologicals brought them out in numbers. During Western films, "when a kissing scene is shown the ladies turn their heads away."[24]

American films usually appeared in India eighteen months after release, although some came much sooner. The most popular film of the decade appears to have been *The Thief of Bagdad,* starring Douglas Fairbanks. Exhibitors almost never saw films before booking them. Distributors said this was not "blind booking" because the exhibitor was told the titles, and could get information from trade papers. Outright purchase of prints—pirated, in some cases—persisted among some traveling cinemas, but others were renting films at Rs. 50 per night.[25] A city theatre would pay a much larger "fixed hire" or there might be a percentage arrangement, if the exhibitor was trusted.

Many Indian producers made only three prints of a feature film, for distribution throughout India. Ten prints appeared to be the maximum. The import duty on raw film was a restraining factor.

Shortcomings of Indian films were often mentioned. But what

[20] *Evidence,* I, 564. [21] *Ibid.,* III, 856. [22] *Ibid.,* IV, 351–65.
[23] *Ibid.,* IV, 90. [24] *Ibid.,* I, 572. [25] *Ibid.,* I, 579.

emerged most unmistakably was the growing preference for Indian films in spite of these shortcomings. Even witnesses who did not share the preference conceded that this was clearly the trend.

Q. You mean that ordinary people—we won't call them illiterate, but not belonging to the middle class—you mean to say they do not go to these theatres where foreign films are shown? Is that what you mean by your answer?

A. Yes, they do not go. . . . Formerly they used to go and see fighting or any exciting films, or comic films.

Q. Now that the Indian films are produced, you think the attendance at foreign films of a social nature is falling?

A. Yes.[26]

Clearly the American serials of the early 1920s had made way, in many theatres, for Indian films. Some exhibitors wanted Indian films but could not get them or afford them.

Q. You find it difficult to get Indian films?

A. Yes, the rates are exorbitant.

Q. Have you ever taken Western films?

A. Yes, they are cheaper than Indian films, but they do not attract the same audience as in the case of Indian films.

Q. But you find it difficult to get Indian films?

A. There is a lot of competition.[27]

In 1918, in Bombay, only one theatre had specialized in Indian films. By 1927 more than half of the twenty theatres showed Indian films at least part of the time. Exhibitors catering especially to a European and Westernized clientele—there were nine such theatres in Bombay—generally felt it essential to stick to Western films. One such exhibitor had shifted for only one week to a Phalke film:

The type of people who like Indian pictures, their way of living is quite different and generally they are people who chew betel leaves . . . let me give you an example. I did show an Indian picture at my Western theatre, *Lanka Dahan,* and I made 18,000 rupees in one week. But it ruined my theatre altogether.

Q. You mean you had to disinfect the cinema?

A. I had to disinfect the hall and at the same time I had to convince my audience I had disinfected it. . . . Till that time I went on losing money.[28]

[26] *Ibid.,* I, 21.
[27] *Ibid.,* III, 318.
[28] *Ibid.,* I, 364.

The problems of Indian film producers were illuminated by many witnesses. Several film makers were producing, or trying to produce, a schedule of a dozen productions a year. A six-week production schedule was considered normal for a feature. Bombay considered Rs. 20,000 a proper feature budget, although a few films had cost much more. Some Calcutta and Madras producers felt that Rs. 10,000, or at most Rs. 15,000, was the practical limit. A Bombay company was paying actors from Rs. 30 to Rs. 1,000 per month. The 30-rupee salary was for "a coolie, a super, an extra"; average actors got Rs. 200–250 per month. A normal star salary was Rs. 600–800 but a few received more.[29] In Bombay producers were already beginning to consider Punjabis the most suitable physical specimens for film acting.[30] Stars were rapidly becoming idols. One woman star, Sultana, received baskets of fruit from distant admirers. In Calcutta a few ladies "of the better classes" had taken part in films—we shall meet some of them later—but some producers drew on women from the "prostitute and dancing-girl class," who had apparently lost their early reluctance toward the cinematograph. The committee, concerned about the well-being of the industry, pursued this matter at every stop:

> Chairman. Do you think that the present conditions in your studio are satisfactory, sufficient to attract respectable actors and actresses?
>
> A. Oh yes, we are catering for respectable actors and actresses.
>
> Q. I mean what arrangements are made for housing them?
>
> A. We keep the respectable characters in separate rooms and they are quite aloof from the others.[31]

Most production was being done by closely knit companies. Each had its own laboratory. Almost all shooting was done outdoors, but a few producers were building, or planning to build, glass-roofed studios. Indian production of topicals was declining, but Pathé Gazette and International Newsreel were shown in a number of theatres. The Madan organization was an occasional producer of topicals, as well as related items: "We have always got a set of cameramen and if we get any orders from Rajahs or Maharajahs to film any function, we can undertake that kind of work too."[32] An un-

[29] *Ibid.*, I, 182–83.
[30] *Ibid.*, II, 13.
[31] *Ibid.*, I, 165.
[32] *Ibid.*, II, 851.

usual, self-taught cameraman in Madras sent reels of film to Fox and International Newsreel in New York City and was paid $2 per foot, undeveloped, for whatever was accepted. He told the committee he had sold 3,000 feet in a four-year period—mainly of festivals and parades.[33] But in general, activity of this sort was sporadic, and what there was of it was beginning to be watched with nervousness by the censors. A newsreel item showing a Sikh procession was banned by the Calcutta Commissioner of Police for fear it would offend Muslims.[34]

In the primitive laboratories maintained by various producers, conditions were often "appalling." Heat was the great enemy of technical standards. A laboratory man explained: "You have got to try and harden your film and it gets nothing but a mass of jelly on the celluloid and the least touch will scratch it. I have had the experience of it washing right off the celluloid."[35]

The relation between film industry and press received attention and produced interesting revelations. The committee noted that newspapers included "critiques" of foreign films more often than of Indian films. Wouldn't searching criticism help to raise the standards of Indian films? Answering this question, a Bombay editor explained:

> If I may frankly confess to you, all newspapers get critique paragraphs typewritten from the exhibitors themselves. That is my frank confession.
>
> Q. In the case of foreign films they get it from the foreign producers, ready made?
>
> A. Ready made, cut and dry, only to be sent down to the printer.

Later he was asked:

> Q. Supposing you criticise a picture honestly?
>
> A. Our trade is so closely interwoven with the interests of the producers and exhibitors that we cannot possibly think of doing so.[36]

A Calcutta journalist told a similar story, while at the same time paying tribute to the *Statesman* for a degree of independence that was apparently remarkable:

> What takes place usually is that for ordinary films the press do not even

[33] *Ibid.*, II, 728. [34] *Ibid.*, II, 1070–71.
[35] *Ibid.*, II, 728. [36] *Ibid.*, I, 495–97.

care to send down a representative. Indeed the reviews which appear sometimes are sent in by the exhibitor. The *Statesman* is about the only paper that cuts it down. The other papers having more space, sometimes it appears as it is sent in.

Q. The advertising revenue does not tie their hands?

A. It does in a way for small papers. But the *Statesman* does not care a jot. Sometimes when Madans have taken a full page they have only received three lines. They have kicked up a row.[37]

Kissing, communalism, motor car dacoity

At each stop in the committee's travels, the workings of censorship received major attention. Under legislation of 1918—the Indian Cinematograph Act—and amendments of 1919 and 1920 the control of cinemas and the censorship of films had been made provincial "reserved" subjects, and placed under police jurisdiction. In Bombay, Calcutta, and Madras, boards of censors had been set up in 1920 to assist the Commissioner of Police in this censorship. A Punjab board had been organized in 1927. Each of these boards could license a film for showing throughout India and also, at any time, "uncertify" it. A film could also be uncertified at any time for any city by its Police Commissioner or for any province by provincial authority.[38] Thus *Orphans of the Storm,* with the Gish sisters, was uncertified in various areas when police found Indian audiences too responsive to the revolutionary scenes; *Razia Begum,* the Indian film about a Hindu-Muslim palace romance—which had angered the Nizam of Hyderabad—was uncertified in various areas after Muslim protests.[39]

The make-up of the boards, and their procedure, reflected prevailing tensions and ways of coping with them. The Calcutta board, for example, had a Hindu member, a Muslim member, a British military member, a British woman member, and others—each representing, in effect, a constituency. British members had a majority. The president, as in the case of each of these boards, was the Commissioner of Police, who was British. In practice, the work of the

[37] *Ibid.,* II, 1080–81.
[38] *Report of the Indian Cinematograph Committee,* pp. 105–9.
[39] *Evidence,* IV, 368–98.

board was largely done by two paid inspectors, one British and one Indian. Every film was seen by an inspector. The board generally certified a film on the basis of his recommendation. If he foresaw possible difficulties, the secretary of the board asked one or two of its members to have a look at the film. This was not done in rotation.

For instance, if the film is such as is likely to be objected to as being offensive to Mohammedans he would naturally put the Mohammedan member to report on it. If it is a film which is likely to be objected to from the point of view of the military, he would put the military representative on it. In the case of a film which is likely to affect women or children the lady representative would be on it.[40]

All this was in accord with accepted practice. It seemed, no doubt, an essential procedure for avoiding trouble on tense issues. At the same time, it had the effect of giving the issues a built-in status. Each board member acquired authority only from the specialized sensitivities he was expected to have. He tended to become a prisoner of his constituency. A group looked to its representative on the board to veto the objectionable. The Muslim board member was thus constantly faced with the choice of either approving a film or defending it thenceforth against the objections of the most sensitive Muslims. The military member, the woman member, the Hindu member—each was placed in a similar position.[41]

The witnesses who appeared before the Indian Cinematograph Committee, and who represented various fields of interest, overwhelmingly favored strong censorship for India. Even people who said they generally disapproved of censorship maintained that it was essential for India. The feeling that India represented a special problem in this respect was, however, explained on widely varying grounds. Some said it was essential because of Hindu-Muslim friction. A member of the Punjab board was asked:

Q. Of course in this province history requires very careful handling?
A. Oh yes.

[40] *Ibid.*, II, 1045.

[41] A college professor on the Calcutta board, not representing an especially touchy constituency, said: "My services were requisitioned not more than half a dozen times in two years." *Ibid.*

Q. Modern history will perhaps have to be avoided?
A. It will have to be for the present.
Q. With the present communal tension?
A. If you want Indian history more modern than 1,000 A.D., it would be difficult to handle the subject.[42]

Others considered censorship essential because, they said, foreign films were encouraging crime in India. A recent rise in robberies involving cars—"motor car dacoity"—was mentioned by several witnesses. All considered these robberies to be due to the cinematograph, although the chairman of the committee at one point suggested that they might also be due "to the advent of the motor car."[43]

Still other people considered censorship essential in India because the hugging and kissing in foreign films were "demoralizing" Indian youth and threatening Indian custom and tradition. Many witnesses favored the strictest censorship of love scenes. Complete elimination of all kissing sequences was recommended by some. One witness felt it would be enough to eliminate close-ups. "Avoid close-ups as much as possible. That is the main thing."[44]

Finally there was the view, held by European witnesses especially, that films were bringing Western society into contempt and undermining Indian respect for Western women.

While the need for censorship was thus asserted on various grounds, political necessity was almost never mentioned among them. Yet when instances of censorship were examined, political reasons loomed large. While the other arguments for censorship were advanced earnestly and with evident sincerity, it is notable that all these arguments, coming from Indians and non-Indians alike, laid the basis for a strict political censorship.

The paid inspectors, before recommending a license, often required the producer or distributor to make specific cuts or changes. The cuts often had to do with subtitles. The committee was curious about a number of these cuts, which represented a variety of problems. Calcutta's British inspector, who considered Kalidasa a writer of books rather than of plays, was asked to explain a cut in an imported film:

[42] *Ibid.*, II, 2. [43] *Ibid.*, I, 80. [44] *Ibid.*, II, 925.

Q. Again look at No. 8070, *The Impossible Mrs. Bellew.* In your remarks you say, "Omit the subtitle, 'Madame, you are magnificent, your figure is fair as your face is beautiful.' " What is wrong with it?

A. It was a direct reference to some essential features of the lady.

Q. Don't you pay compliments to a lady?

A. I think it was somewhat an offensive remark to make.

Q. Sometimes they are flattered too. For instance, our Kalidasa has put in some such words to Dushyanta in addressing Shakuntala?

A. But I would not draw an exact parallel between a book and a picture on the screen. You would agree that a picture shown on the screen is much more striking and will appeal to a much wider audience than it can possibly do in the case of a book.[45]

The committee pried into a number of subtitle cuts made in Bombay and Calcutta, both in Indian and in foreign films. Many cuts involved political reasons. The extent and nature of them seemed to surprise some committee members.

Q. Then again in Reel VIII the words "in freedom" have been cut from the title. "My sons! Die in freedom rather than living in Shivaji's service." Why was it done?

Secretary of the Board: It was thought to have some political significance in it.

Q. Was it done by the Board or by the Inspector?

A. It was done by the Inspector with the concurrence of the producer.[46]

Q. Then again take 7640, page 33, *Bright Shawl,* where you say . . . "Omit the subtitle, 'And my poor brother's only sin was to love his native land.' " What is wrong with it?

A. It is impossible, I think, to judge these things apart from the context simply from the brief notes put down here.[47]

Q. You say further, "Omit the subtitle, 'To us of Royalty can anything be sweeter than the smell of a dead traitor?' " What is wrong with it?

A. That is a matter of opinion; we would not like to have it applied to our own Royal Family.[48]

Q. You say, "Omit the title, 'I have revised the civil list—increasing all our salaries by one-third, etc.' " I don't see anything wrong in it. Does this indicate the nervousness of people who benefited under the Lee concessions?

A. I don't see how you could draw any conclusions from these brief notes.[49]

[45] *Ibid.*, II, 556.
[46] *Ibid.*, I, 101. [47] *Ibid.*, II, 573.
[48] *Ibid.*, II, 555. [49] *Ibid.*, II, 573.

Additional cuts and revisions that interested the committee:

Fortune's Mask. . . . Omit in Part I . . . "He is getting popular—have him investigated."[50]

Title . . . "Dreamed of a day when the government would be a government of the people, by the people, for the people" . . . ordered to be substituted by, "Dreamed of a day when peace and contentment would prevail in the land."[51]

For the subtitle, "We will hold a mass meeting in the square and force the President to declare peace or war," substitute, "We will hold a mass meeting in the square."[52]

For the subtitle, "But that is murder, they are our own people," substitute, "Must I obey your orders, sir?"[53]

Some committee members wondered about the effect of such interference on the producer and on the development of the motion picture. A producer was questioned about this:

Q. You are asked to omit, "Oh God, I have always been a man of peace. But the ways of peace seem to have gone wrong. Please guide me." . . . Do you think an English director or an American director would stand any such treatment?

A. Well, I think he would resent it.[54]

But most producers said little by way of protest. In fact, seldom during the weeks of hearings did any witness oppose censorship on principle. The record does contain one spirited statement of this sort, by A. Venkatarama Iyer, B.A., B.L., of Madurai:

I think every member of this committee believes in the freedom of speech and freedom of opinion. I believe that all must have read John Milton's *Aereopagitica*. I believe also that British citizenship is a thing founded upon liberty. I think that classical works are characteristically great because there is freedom of expression and boldness of conception. Fetters, even though they are made of gold, are still fetters. Censorship is cold, critical, routinelike, tyrannous, and inspires fear in the budding genius to express himself. The business of the censor is more to prohibit rather than appreciate a work of art. The very name savours of a sickening restriction, and it is the hand of death if it touches a work of art.[55]

But opposite views were urged with equal vigor:

Unduly interfere with the artistic and inspirational development? This is bosh! There is neither art nor inspiration in such pictures. They are gross and vulgar.[56]

[50] *Ibid.*, II, 1059. [51] *Ibid.*, II, 933. [52] *Ibid.*, II, 1060.
[53] *Ibid.* [54] *Ibid.*, I, 191. [55] *Ibid.*, IV, 244. [56] *Ibid.*, I, 385.

In May of 1928 the committee submitted its report. As instructed, it made recommendations on (1) the adequacy of censorship and (2) imperial preference. On the matter of censorship it took a calm tone, expressing the opinion that Indian youth was not being demoralized and that many of the alarms about the impact of film in India were exaggerated. It emphasized that many of the expressions of alarm had originated outside the country, and suggested that they had come to a large extent from people motivated by their own special interests, and perhaps not fully in touch with the facts.

As to the adequacy of censorship, the committee expressed itself as satisfied. It gave cautious support to the view that "too much tenderness is bestowed on communal, racial, political and even colour considerations," and suggested that "over-much tenderness to frivolous objections is more likely to encourage dissension." However, it recognized that "the vast majority of witnesses . . . consider that censorship is certainly necessary in India," and it expressed its concurrence with this view. It suggested that a central board of censors could help to develop some uniformity in the standards of censorship, while still leaving a good deal of authority in local hands.

On the matter of "Empire films" the committee was forthright.

> If too much exhibition of American films in the country is a danger to the national interest, too much exhibition of other Western films is as much a danger. . . . The British social drama is as much an enigma to the average Indian audience as the American. In fact very few Indians can distinguish American manners and customs from British manners and customs; very few Indians can distinguish an American, German, or Frenchman from an Englishman or Scotchman. If the cinema therefore has any influence on the habits, lives and outlook of the people all Western films are likely to have more or less the same kind of effect upon the people of this country.[57]

With these words the Indian Cinematograph Committee thrust aside the idea of imperial preference.

But it went further. It said the important thing was to nurture the Indian film. For this purpose it urged various measures, including a cinema department under the Indian Ministry of Commerce

[57] *Report of the Indian Cinematograph Committee,* pp. 99–100.

to look after the interests of the film industry, a government film library to utilize the educational value of film, a governmental film finance fund to aid producers with loans, and a government plan to encourage the building of cinemas.

The committee went still further, and recommended the abolition of all import duty on raw film. "That the raw material of an industry should be free of duty is almost axiomatic."[58]

The report included still another proposal, although it was at this point that the British members parted company with the Indian members and issued their Minute of Dissent. The additional proposal was a modified quota plan requiring Indian theatres, with some exceptions, to show a minimum proportion of *Indian* films. "That the best theatres in her own country should not be open to her own productions is a reproach which must be removed."[59]

While this proposal clearly echoed the spirit of the British quota plan of 1927, the British members now dissented. Their Minute of Dissent pointed out that, strictly speaking, the committee was evenly divided, 3–3, on this matter, and that even though the other group included the committee chairman, its proposal could not properly be called a majority recommendation, nor properly included in the report. It also frowned, to some extent, on the proposed financial support to Indian producers.

But all this hardly mattered, for the Government of India completely ignored the recommendations of the Indian Cinematograph Committee. Not one of them was enacted into law. This was perhaps not surprising, since the committee had rejected the very premises on which its existence had been based. However, some of its ideas would be revived in a later day, by a Film Enquiry Committee of independent India.

As to residual effects of the Indian Cinematograph Committee, a few may be suggested. Although the committee had expressed warnings about censorship, it had on the whole confirmed, and left undisturbed, a strict censorship system. Rigorous film censorship thus continued to be an Indian habit, a habit that would not readily be put aside by a government of independent India.

Now that imperial preference had been rejected for the world of

[58] *Ibid.*, p. 75. [59] *Ibid.*

film, it was also clear that the dominance of American films among film imports would persist. The story of film in India would continue to be, for many years, a story of American and Indian films. And since an Indian quota and government assistance were likewise dead issues, the Indian film would have to make its way against its formidable rival without quota or subsidy.

The failure of the Government of India to respond to the recommendations of the Indian Cinematograph Committee may have had another reason besides that we have mentioned. The committee had been appointed on October 6, 1927. That was also the day on which *The Jazz Singer,* the world's first talking feature, had its premiere in New York City. Its reception signaled the end of an era.

Thus the film world which the committee studied so assiduously was already marked for sweeping changes. By the time the committee recommendations were written, detailed reports on *The Jazz Singer* and its impact on American audiences were appearing in Indian papers. Throughout 1928 the film trade press informed Indian producers of Hollywood's hectic scramble toward the new era. In 1929 *The Melody of Love,* a Universal Pictures production, became the first sound feature to be shown in India. Indian producers read, saw, heard—and knew that it was all inevitable. To many, it must have seemed like a pronouncement of doom.

Discord of Tongues

The market of the Indian producer, up to this hour, had been an area inhabited by several hundred million people. Burma and Ceylon were being administered as part of India and within this area no barriers—political, economic, or linguistic—had barred the way to Indian films. Occasional successes had gone to Malaya, East Africa, South Africa. But now a film would apparently need a language.

In Bombay, which was leading in production volume, it meant this: Located in the Marathi-speaking area of the country, its producers would naturally make films in that language. If so, they would at the start of the 1930s have a potential market of 21 million people, almost all in the region surrounding Bombay and including Poona, Kolhapur, and other cities.[1] But the films would be incomprehensible in the rest of India, in Burma, and in Ceylon.

In Calcutta, which stood second in production, it meant this: Situated in Bengal, its producers would naturally make films in the Bengali language, which would give them a market area inhabited by some 53 million, largely in the northeastern portion of India.[2] Again, the films would be incomprehensible in most other parts of India, in Burma, and in Ceylon.

In Madras, which had made a hesitant start in film production, it meant this: Situated in the Tamil-speaking area of the country, its producers would logically make films in that language, which would give them a potential market area of 20 million people in southern India and some additional millions in Ceylon, Malaya, and Africa. But the Tamil language, of Dravidian descent and unrelated to Marathi, Bengali, and other north Indian tongues, would make the films incomprehensible in most of India.

Would film producers, accustomed to visions of wide, growing markets, now be hemmed into linguistic pockets? Instead of competing in a large area, would they chop it into zones? And could an

[1] Statistics in this passage are from Chatterji, *Languages and the Linguistic Problem,* p. 14. [2] Including areas now in Bangladesh.

SOUND ERA BEGINS:
Alam Ara, 1931
COURTESY NATIONAL FILM ARCHIVE OF INDIA

area of 20 million, or 21 million, or even 53 million inhabitants support a film industry, if costs should rise steeply in the era of sound?

Curiously, none of the major film centers was situated in the largest linguistic zone, comprising the 140 million Hindi-speaking people, mainly in north central India but with additional clusters in other parts. This Hindi market, in view of its size, would clearly be the most important; even so, films in Hindi would not be understood in vast areas, including most of the south.

Also curiously, no major film center was located among the 28 million people speaking Telugu, another of the Dravidian languages. Largely rural and including no metropolitan centers, this area had never generated film enterprise. But it would be an important market, that could not be well served by films in any language except Telugu.

If the large areas speaking Hindi, Bengali, Telugu, Tamil, and Marathi might conceivably support regional production, there would still be the problem of smaller but by no means negligible language pockets: Punjabi, 15 million; Gujarati, 11 million; Kannada, 11 million; Malayalam, 9 million; Assamese, 2 million; Oriya, 2 million; Kashmiri, over 1 million; and others, including several million speaking primitive tribal tongues.

While films in Hindi would obviously acquire importance, this raised a further question: What kind of Hindi? For the Hindi-speaking area offered a linguistic chaos of its own, and had for centuries.

In ancient India, when Sanskrit was the language of courts, the common people already spoke a diversity of Prakrits, which tended to become more diverse. The huge, populous plains of the upper Ganges and Indus and their tributaries, comprising a dense mass of villages and towns, came to represent an extraordinary tangle of regional tongues. Gradually, as a means of communication within this area, a common "bazaar language" emerged during the middle ages, and Hindi was a development of this. In simplest form it is often called Hindustani, as the area that produced it was often called Hindustan.

As Hindi in the last century or so began to develop a literature, it tended to enrich its vocabulary by dipping back into Sanskrit, just as modern European languages coin new words on Latin stems. Thus literary Hindi, administrative Hindi, and, in recent years, radio Hindi have tended to be a Sanskritized Hindi.

Meanwhile the Muslims, having taken hold of the same medieval bazaar language, had amplified it with Persian words, producing a Persianized Hindi known as Urdu. It is the language of Muslims in many parts of India. Hyderabad in south central India, for example, has a substantial cluster speaking this Persianized Hindi.

Thus the film producer was faced with a problem of practical and emotional dimensions: Sanskritized Hindi, Persianized Hindi, lowbrow Hindi—what kind of Hindi?

The dismaying problems facing Indian producers now threw into relief their good fortune of earlier years. It became clear how uniquely blessed they had been. India, along with its disruptive forces of language, religion, and caste, has had other forces making for unity, and it was these the silent film had been able to enlist—while evading the others.

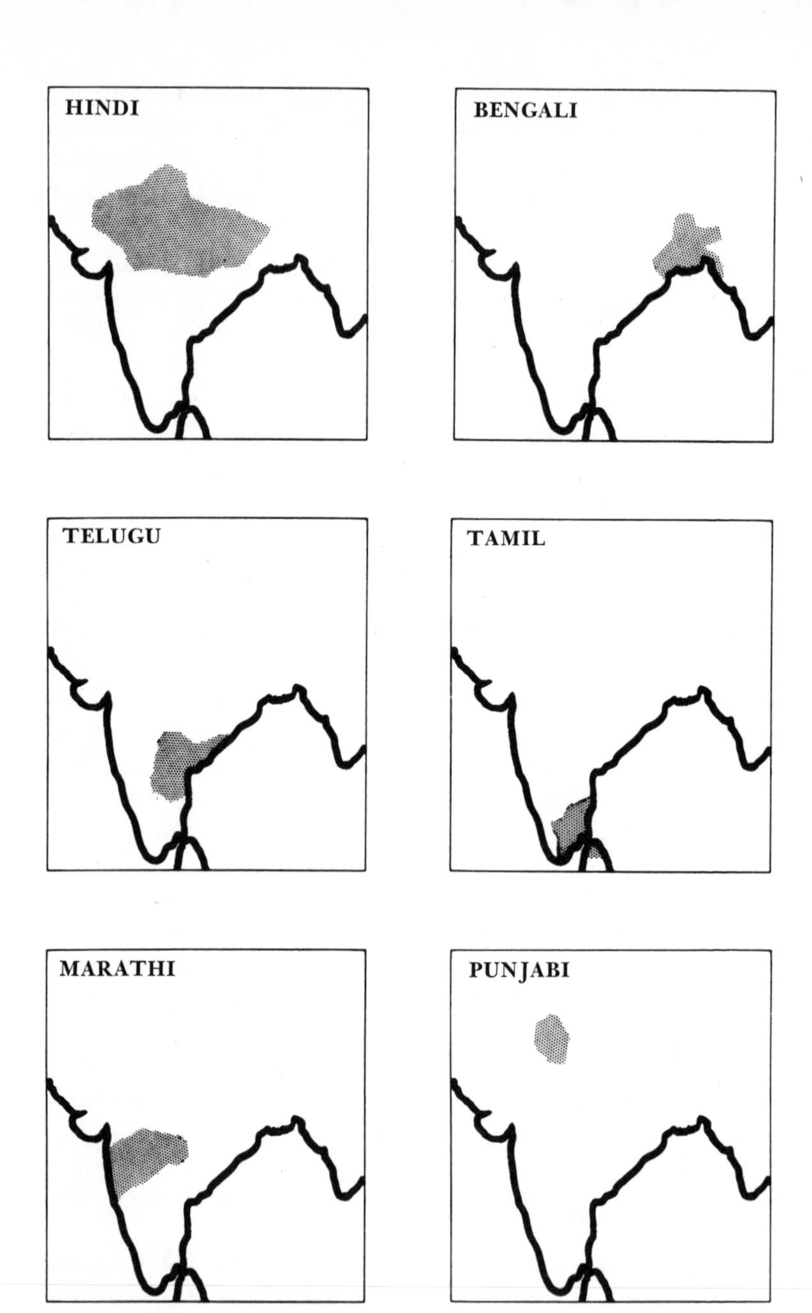

DISCORD OF TONGUES: INDIAN LANGUAGE AREAS

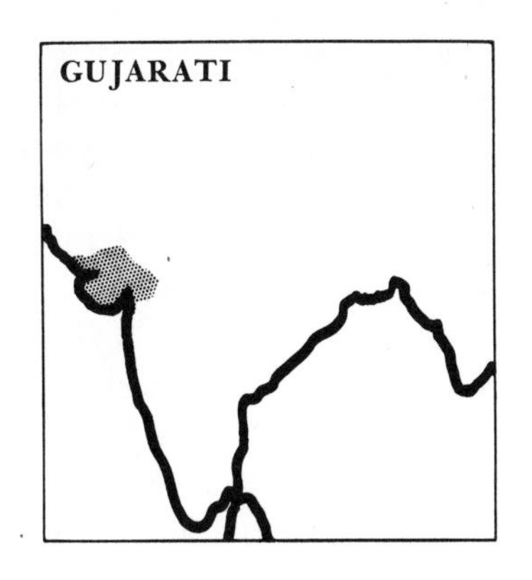
GUJARATI

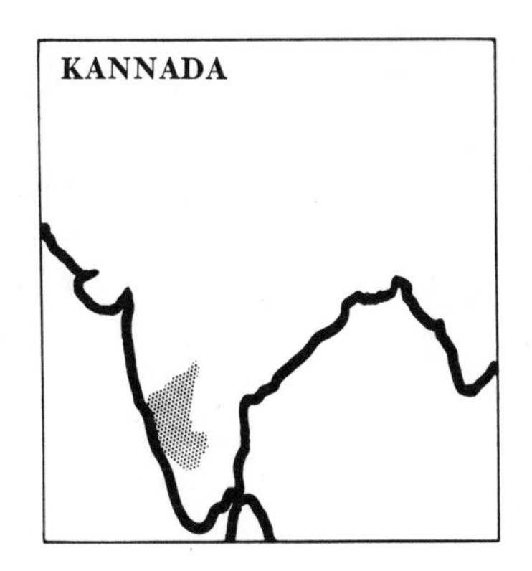
KANNADA

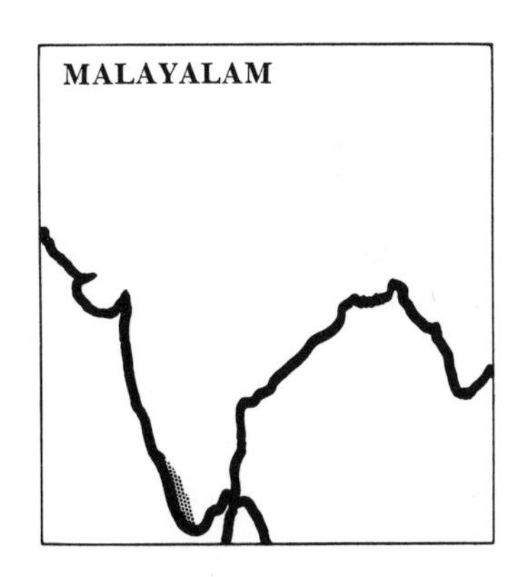
MALAYALAM

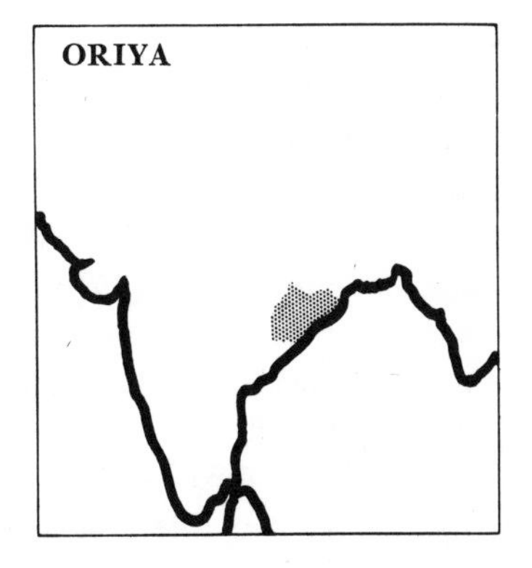
ORIYA

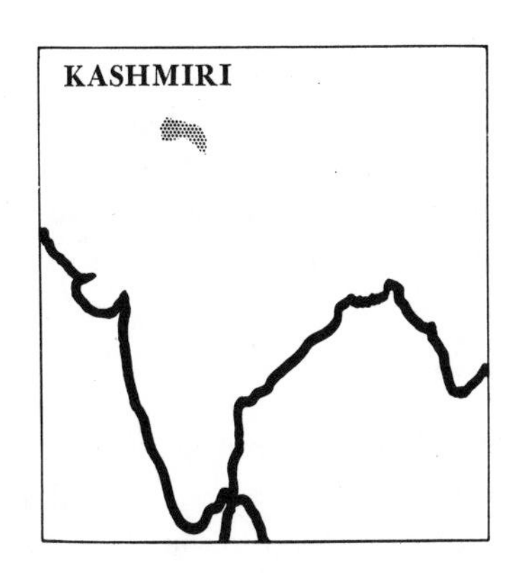
KASHMIRI

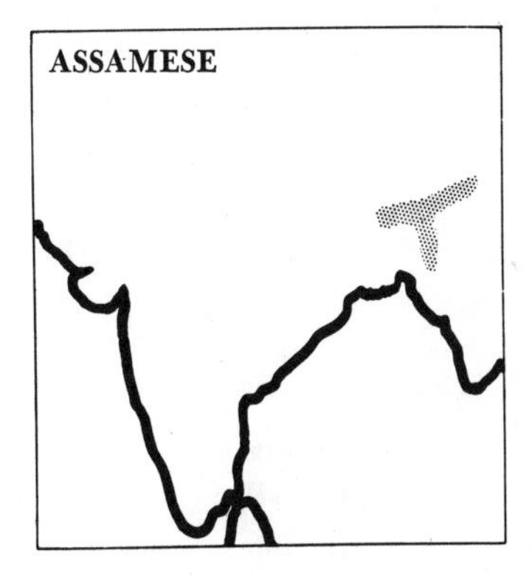
ASSAMESE

An Indian sense of community has been especially fostered by common cultural legacies, among the most remarkable of which have been the great Indian epics, the *Ramayana* and the *Mahabharata*. Like the *Iliad* and *Odyssey,* they are oversized tales of adventure involving gods, heroes, and mortal men, as well as mythical animals. Each has a unifying thread of plot, holding together a vast panorama of people and incidents. Like the Bible, each is also a compendium of folk history, poetry, and wisdom. Told, sung, and acted long before the Christian era, these tales have conquered the waves of invaders that have swept over and fragmented India. The stories found their way eventually, in varying versions, into all the languages of India, and also spread to other regions of South Asia such as Ceylon, Burma, Malaya, Sumatra, Java. Throughout this region, and even beyond it, the characters of the epics appear in song, dance, drama, painting, sculpture. This bond had magnificently served the silent film. But now, for the first time, disruptive forces would come to the fore.

The problems seemed so difficult that, at the first news of the advance of sound, a number of film production units quietly folded. Sound would mean investment in expensive equipment, to be imported at much risk. It would apparently call for a studio. A studio, if soundproofed, would require lighting equipment, seldom used heretofore in India. Sound and artificial lighting would require skills not yet available. Above all this would be the problem of uncertain and restricted markets. Many of the film units that had managed to survive in the silent era had neither resources nor knowledge with which to face the future. We have already mentioned some of those that now passed into oblivion: British Dominion Films, Eastern Film Syndicate, General Pictures Corporation, Associated Films, and others. But the climax of these events was totally unexpected. In 1931 Madan Theatres Ltd. began to come apart at the seams.

Death of a giant

If any company had seemed ready to face the new age, it was Madan Theatres Ltd. After the death of its founder, J. F. Madan, the company had undergone several shuffles at the top, but before long J. J. Madan, third of the five Madan sons, had become managing director. Though disagreements between the brothers caused intermittent difficulties, J. J. Madan soon appeared to be giving effective leadership. Under his management the chain of theatres continued to expand. The company had owned or controlled 51 theatres in 1920; these grew to 85 in 1927 and to 126 in 1931.[3]

Throughout these years there were rumors that Madan Theatres would be sold to an American film company. These rumors, which caused anxious flurries in Indian film circles, had a basis in fact. J. J. Madan made a number of trips to Europe and America and on one of these he negotiated with Carl Laemmle, president of Universal Pictures Corporation. They agreed on sale terms, subject to approval by their respective companies. But when he returned to India, J. J. Madan found his brothers unwilling to approve the negotiated price.[4] Perhaps they thought it could be bettered by further negotiation, but the course of events wiped out any such possibility.

In December of 1927, during the investigations of the Indian Cinematograph Committee, J. J. Madan testified for two full days on behalf of Madan Theatres, and proved a forthright, winning spokesman.[5] He seems to have persuaded committee members that charges of monopoly practices, made by some witnesses against Madan Theatres Ltd., were "preposterous" and that the Madan organization was merely more experienced and alert than its competitors.[6] Soon after his testimony he journeyed again to the United States.

In New York he saw Al Jolson in *The Jazz Singer* and witnessed at firsthand its impact on filmgoers and the film industry. He found the American film world in a frenzy of retooling and reorganization. Hollywood was firing screen writers and commandeering Broadway playwrights. Long-established star contracts were being canceled and new talent wooed. A spirit of panic and plunge gripped many companies. J. J. Madan caught the fever, ordered sound production equipment, and headed back for India. In 1929 Madan Thea-

[3] The 1920 and 1927 figures are based on *Evidence,* II, 844–45; the 1931 figure on an interview with J. J. Madan.

[4] Interview, J. J. Madan.

[5] See his testimony and colloquy with committee members, *Evidence,* II, 835–90.

[6] *Report of the Indian Cinematograph Committee,* pp. 43–45.

tres ushered in the talking picture in India by premiering, at the Elphinstone Picture Palace in Calcutta, Universal's *Melody of Love*. Meanwhile construction of a soundproof Madan studio had been started on the outskirts of Calcutta at Tollygunge—soon to be dubbed "Tollywood." An ambitious sound production schedule was planned.

But not everything went according to plan. That autumn brought the New York stock market crash, spectacular overture to a long, world-wide depression. Any lingering thoughts of a sale to Universal evaporated. Universal had other problems.

So did Madan Theatres. The studio was completed on schedule and the production plans launched. But their success would depend on the conversion of the chain of theatres to sound. The Madan brothers, involved in innumerable enterprises, many hinging on import and export, found their cash position threatened by the growing world paralysis. The conversion of the theatres began to loom as a major obstacle. And the theatres were reporting declining revenue.

In part this decline may have reflected an attendance drop. But in part it reflected something quite different. The Madan family began to receive evidence of a deeply disturbing problem. With the apparent collaboration of paid inspectors, attendance figures at a number of Madan theatres were being misreported by paid managers. At an apparently growing number of theatres, profits were being syphoned away. In a theatre empire as scattered as the Madan chain—over India, Burma, Ceylon—in areas so forbidding to efficient supervision, how could the company hope to defeat this threat?

As suddenly as J. F. Madan had plunged into tent "bioscope" showings in 1902 and launched picture palaces a few years later, J. J. Madan began to sell theatres in 1931. Once the decision was made, properties were disposed of in rapid order, largely one by one. In less than two years only one Madan theatre remained—the Regal in Calcutta. Presently the old theatrical company at the Corinthian, seriously foundering, was liquidated and the theatre converted to film. As the Opera Cinema, it brought the Madan cinema holdings back to two theatres.[7]

[7] Interview, J. J. Madan.

Meanwhile Madan Theatres had become the first Indian producing organization to release sound films. It made a number of short items and on March 14, 1931, put on a program of 31 such films. They included a hymn chanted in Sanskrit by "lady worshippers at the temple of Siva," a girls' school chorus singing a Tagore song, a dance by "the Corinthian girls," a scene from a Hindi (Urdu) play, a recitation from Kalidasa, a speech by Indian Nobel Prize physicist Sir C. V. Raman, and other items including a comic song.[8] On that same day a Bombay producer, the Imperial Film Company, became the first to release a sound feature—*Alam Ara* (Beauty of the World). A number of Madan feature films, some in Bengali and some in Hindi, followed this in rapid succession: eight Madan features were released in 1931, sixteen in 1932.[9] But the company deflation had not ended. The following year brought a slowdown in its production. The studio was offered for rent to other producers; finally it was sold. Except for the two surviving cinemas, the Madan film empire passed out of existence.

Today its rise and fall are an almost forgotten story. Its main memorial is a street in the heart of Calcutta, renamed in honor of the founder of Madan Theatres Ltd. shortly after his death in 1923. The Bengal Motion Picture Association would have offices on Madan Street. Its fading splendor would also become home to the Anti-Rowdy Section of the Calcutta Police Department.

Sound statistics

The disintegration of the one sizable organization in the Indian film world was ominous news. Even so, the sound era got started.

The statistical story of the Indian sound film in its earliest years may be briefly summarized. It was, in large part, a story of new units, in which individuals from older companies were brought together by new capital.

The Bombay producer who made the first talking feature, *Alam*

[8] *Filmland,* March 21, 1931.
[9] *Indian Talkie, 1931–56,* list, pp. i, xviii. In this list Hindi and Urdu films are both counted as "Hindi" films. Similarly, government statistics now lump Hindi, Urdu, and Hindustani together. We shall follow the same practice, using the term "Hindi."

Ara—in the Hindi language—was Ardeshir M. Irani. Born in 1885, he had started out in his family's musical instruments business, grown restless, gone into distribution of foreign films, and finally joined with tent showman Abdulally Esoofally in buying the Alexandra Cinema in 1914 and building the Majestic Cinema four years later. Exhibition profits edged the partners into production. After involvement in several other companies they launched the Imperial Film Company in 1926,[10] and built a studio for it. In 1931 this company won the sound race among Bombay producers. The equipment Irani obtained from the United States was virtually "junk"[11] but somehow, via its single-system process,[12] he completed *Alam Ara*. The film has never been described as an artistic triumph and no one seems to have preserved even a fragment of it. But its impact was astonishing. The Majestic theatre was besieged. Tickets disappeared into the black market. "Police aid had to be summoned to control the crowds. . . . Four-anna tickets were quoted at Rs. 4 and Rs. 5."[13] Later, units went on tour with the film, taking sound projection equipment with them, and everywhere drew surging crowds.

That same year 22 other Hindi films appeared, and all seem to have made money. Also in 1931, three films in Bengali, one in Tamil, one in Telugu, appeared in their respective language areas. The year 1932 brought eight films in Marathi, two in Gujarati. In 1933, 75 Hindi features were made; production in other languages was also growing.[14] Film after film appears to have had a tumultuous reception. Virtually all the films appear to have earned back their cost. In the 1930s, as one producer recalls wistfully, "almost all films made money."[15]

By 1933 trepidation over the coming of sound had given way to unbounded optimism. That year the compiler of *Who Is Who in*

[10] *Indian Talkie, 1931–56,* p. 122.
[11] Interview, Irani.
[12] In this system, later used mainly for newsreels, sound goes directly onto the picture negative. In the more versatile double system, picture and sound are kept separate for flexibility in editing, to be combined in the laboratory as one of the final steps of the production process.
[13] "Half a Century in Exhibition Line," in *Indian Talkie, 1931–56,* p. 121.
[14] *Indian Talkie,* 1931–56, list, pp. i–xxvii.
[15] Interview, B. N. Reddi.

Indian Filmland, in a jubilant preface, gave expression to the mood:

> What with scanty resources, stepmotherly Government aid, with keen competition from privileged foreign films, with few technically qualified men, with no interested capitalists, with less interested fans, with actors and actresses scarcely able to spell their names (for it was thought a disgrace by society people to be associated with the screen), with no market excepting India, with censuring censors, with discouragement to the right, cheap sneers to the left, despair in front, and criticism from behind, the Indian Film Industry, thank God, has marched on and on to the field of victory, battling against a thousand other misfortunes. Has she not made a giant stride?[16]

What had made possible this sudden reversal of fortune? How had such startling success been won?

Undoubtedly several factors had been at work. The status that had suddenly been conferred by film on the vernacular tongues, in a land in which foreign languages had for a thousand years dominated the councils and pleasures of the mighty, was a powerful influence. There was also the fact that sound had granted the Indian producer a "natural protection." Though facing new problems, he now had markets which foreign competitors would find difficult to penetrate. The protection which the Government of India had declined to give him through a quota system had now been conferred by the coming of the spoken word. But along with these factors, an even more potent force had been at work.

Alam Ara included about a dozen songs. Another early Hindi film is said to have had about forty songs.[17] An early Tamil film is said to have had over sixty songs.[18] All the sound films produced in India in these early years had a profusion of songs. Most also had dances. Advertisements described some of these films as "all-talking, all-singing, all-dancing" features. The Indian sound film, unlike the sound films of any other land, had from its first moment seized *exclusively* on music-drama forms. In doing so, the film had tapped a powerful current, one that had given it an extraordinary new impetus. It was a current that went back some two thousand years.

[16] *Who Is Who in Indian Filmland,* p. 1.
[17] *Indian Talkie, 1931–56,* p. 83.
[18] Interview, T. R. Sundaram.

In ancient India, in the Golden Age of Sanskrit theatre, the idea of drama was already inseparably linked with song, dance, and music. In fact, we are told that Sanskrit and some of its derivative languages had no separate terms for "drama" and "dance," and that the notion of drama as a separate entity, independent of the other elements, is still strange and "disconcerting" to many Asians.[19]

The dramatic practices followed by Sudraka (*ca.* 300 A.D.), Kalidasa (*ca.* 400 A.D.), and other Sanskrit playwrights were codified in the famous treatise *Natyasastra,* ascribed to the sage Bharata. This work has played, in Indian dramatic theory, a role similar to that played in the Western world by Aristotle's *Poetics.*

Bharata, stressing the importance of music to drama, said: "Instruments are the very bed of a performance." A Sanskrit dramatist, justifying the number of songs in his plays, is said to have explained that they "delight the hearts of the audience and establish the emotional continuity."[20]

After Kalidasa the Sanskrit theatre went into a long decline and, after 1000 A.D., virtually expired. Its death was hastened by the various waves of Muslim invaders who swept over and ruled large parts of India during the following centuries. The Muslims had no theatrical heritage and at this time considered drama a sacrilegious activity. In the Muslim era court patronage of drama ceased throughout most of India. Sanskrit plays continued to be written after ancient models but as an exercise of scholars. Sanskrit, displaced as the language of courts, fell further into disuse, and the Indian classical heritage became inaccessible to all but a few. The theatre, for centuries, ceased to exist.

In the nineteenth century, under the British, Indian drama underwent a rebirth. It flourished first in the form of private family theatres maintained in the large joint-family homes of educated Indian families, especially in Calcutta. It was in such a private theatre that Rabindranath Tagore had years of experience as writer and performer before emerging, a seasoned artist, on the public

[19] Bowers, *Theatre in the East,* pp. 9–10.
[20] Quoted *ibid.,* pp. 24–25.

stage.[21] Both in the private theatres, which flourished from the 1830s, and the public theatres, which began in the 1870s, the new Indian drama started by adapting and imitating European models. But almost at once there was a reversion to ancient usage. As Indian theatre activity grew in various cities, drama, song, and dance once more became inseparable entities.

To Faubion Bowers, author of the fine, panoramic *Theatre in the East,* there is something almost mystical about this. The canons of theatrical art as practiced by Kalidasa had somehow, he feels, "remained as a kind of invisible law."[22] But other explanations have been offered.

In India the drama, much in the manner of the drama of ancient Greece, had originally developed from dances performed at religious festivals. As these added elements of narrative and dialogue, they became a kind of folk drama. The *jatra,* a form of folk drama long popular in Bengal and surrounding areas, apparently stems from this ancient period. *Jatra* means festival, but in Bengali the word also came to be applied to plays performed at festivals. Songs were always a central feature of the *jatra.*

While Sanskrit drama became an ornament of the courts, the *jatras* continued to entertain the common people. When drama was banished from the courts, the *jatras* continued in the villages. Traveling players, or *jatrawalas,* continued to journey from village to village, from festival to festival. The theatre was dead but the *jatra,* which needed no stage, lived on in unbroken continuity. Without patronage and generally looked down on by the educated, the *jatras* were not preserved in literary form but maintained a vigorous life. Ironically, dramas "were written but seldom acted, while *jatras,* which were acted publicly, were not written down till the end of the nineteenth century."[23]

There is no doubt that the *jatras,* crude and naïve as they may have been, had a powerful hold over large audiences. Some *jatrawalas* were idolized. The *jatra* also became at times an instrument of religious and social reform. The Vaishnava movement, which

[21] Sen, "Bengali Drama and Stage," in *Indian Drama,* pp. 53–54.
[22] Bowers, *Theatre in the East,* p. 18.
[23] Sen, "Bengali Drama and Stage," in *Indian Drama,* p. 46.

rose in the sixteenth century and had the power of love as its central theme, made vigorous use of the *jatra* and other forms of folk music-drama for the lyrical propagation of its teachings.[24] And even in the first decade of the twentieth century Mukunda Das, a Bengali *jatrawala,* is said to have used the medium to urge the reform of Hindu society.[25]

Corresponding to the jatras of Bengal and adjoining areas, other forms of musical folk drama persisted through the centuries in other parts of India. There were the *ojapali* of Assam, the *jashn* of Kashmir, the *kathakali* of Kerala, the *leela* of Orissa, the *swang* of Punjab.[26] When a new Indian theatre began to develop in the nineteenth century, these folk-drama forms exerted an immediate influence: a vast tradition of song and dance was available to the new theatre. When the sound film appeared, this same reservoir pressed strongly upon it.

Thus the Indian sound film of 1931 was not only the heir of the silent film; it also inherited something more powerful and broad-based. Into the new medium came a river of music, that had flowed through unbroken millennia of dramatic tradition.

While this strengthened the film, it also had other effects. It meant an almost mortal blow to the *jatras* and other kinds of folk drama. The itinerant cinemas shouldered aside the traveling *jatrawalas* and took their place in the hearts of the people. As for the reborn theatre, the sound film almost wiped it out with one brush of its hand. Only gradually has it struggled back to a show of life.

There were other problems too. As the film appropriated folk song and dance to its purposes, it changed them. In their new environment they began, quite naturally, to respond to new influences. The songs were transformed through new instrumentation and new —sometimes Western—rhythms. Musicologists, just beginning to discover this same folk music, and to prize the way a song was sung in Assam in 1875 or in Orissa in 1892, howled in fury. "Hybrid music!" they cried—and are still crying, in protest against "film music." But that is a story for later pages.

[24] Debi, "Assamese Drama," *ibid.,* pp. 37–38.
[25] Sen, "Bengali Drama and Stage," *ibid.,* p. 54.
[26] These and other folk-drama forms are discussed in *Indian Drama,* pp. 36, 75, 79, 95, 97.

In 1931 and 1932, at what seemed a dark moment in Indian film history, song and dance—in part derived from a tradition of folk music-drama—played an important role in winning for the sound film an instant and widening acceptance. "With the coming of the talkies," wrote a contemporary observer, "the Indian motion picture came into its own as a definite and distinctive piece of creation. This was achieved by music." He also observed that this same music might, for a time, tend to block the Indian film from Western markets, and this proved to be a perceptive prophecy.[27] It was also noted by observers that the obsession with music was a hazard to script values. A film periodical commented: "Cases of singing before drawing a sword for a fight are not uncommon."[28] In the Indian film world writers would have problems.

We have mentioned that the early sound era was dominated by new production units. Three such units were to exert special leadership. So important was their role that we must examine each in detail.

[27] Desai, "Overseas Market for Indian Films," in *Indian Cinematograph Year Book, 1938,* pp. 291–93.

[28] *Journal of the Motion Picture Society of India,* June, 1937.

Studio

One new company emerged in Tollygunge, Calcutta, in 1930. As Madan Theatres Ltd. disintegrated in the following years, New Theatres Ltd. rose rapidly.

Its creator was a young man still in his twenties, Birendra Nath Sircar, son of Sir N. N. Sircar, Advocate-General of Bengal. Born in 1901 in Bhagalpur, he was sent to England for part of his education, studying engineering at the University of London. He returned to India to pursue a career as engineer and builder. But one of his first building tasks was a cinema, and this contact with the film world proved a turning point. The young man decided to build a cinema for himself, and then became involved in two silent film ventures, of no special distinction. But he was learning his way.

Always soft-spoken, described as "the most well-behaved gentleman in the film world,"[1] B. N. Sircar was a contrast to many around him. He would quietly and carefully study a problem, then decide and proceed. Unlike most film leaders, he seemed to have no consuming ambition to be performer or director. Putting the right pieces together was his specialty. His pleasure was to give a good director the budget he needed, and let him go ahead without interference. B. N. Sircar was the first example in Indian film of the creative "executive producer."

Having decided to form a company to produce sound films, he moved ahead with speed and precision. No doubt his family connections were helpful. As the son of the Advocate-General of Bengal, he could apparently raise money at will. In fact, the chief investor in New Theatres Ltd. was said to be the Advocate-General.

By 1931 B. N. Sircar had built and equipped a first-class studio and laboratory and gathered around him varied talents. In choice of personnel he most clearly showed the quality of his leadership.

[1] *Who Is Who in Indian Filmland,* p. 25.

DEBAKI BOSE

Some of those he chose are already familiar to us. When we last saw Dhiren Ganguly, the satiric comedian and director of *England Returned,* he was trying to recover from his encounter with the Nizam of Hyderabad and to form a new company in Calcutta. Several years of effort resulted in British Dominion Films, which finally went into action early in 1929,[2] but collapsed with the Indian switch to sound. After brief association with other ventures, Ganguly threw in his lot with Sircar and under the New Theatres banner directed several comedies before going elsewhere. The most successful of these, made in Bengali and Hindi versions, was *Excuse Me, Sir.*[3]

Another of those recruited by Sircar was Debaki Bose. This intense young nationalist, after writing and acting in *Flames of Flesh,* had gone on to directing—for British Dominion and others.

[2] *Amrita Bazar Patrika,* February 10, 1929.

[3] Ganguly has explained its genesis as follows. Literary piracy and brain-picking were a constant problem. Therefore, when asked what his next film would be, Ganguly usually answered, "Excuse me, sir," and went on to another subject. When people began to ask when *Excuse Me, Sir* would be finished, he decided he should produce such a film, and did. Interview, Ganguly.

One of the silent films he directed, *Aparadhi* (The Culprit), made for a short-lived company called Barua Pictures, won high critical praise. But it was sound that brought out the real talents of Bose. Steeped in the traditions of the Vaishnava movement and its musical evangelism, he became a specialist in lyric, devotional dramas.

His first talking picture assignment under the New Theatres banner was *Chandidas* (Chandidas), the story of a Vaishnavite poet-saint of the sixteenth century. Congenial to Bose, the topic was also suited to the developing music-drama of the screen. The film was saturated with music. It included songs based on the work of Chandidas and also much background music. Through *Chandidas,* Debaki Bose taught the Indian film world what could be done with background music. He showed that it could take over functions of dialogue and, while reducing dialogue, intensify it.[4]

Chandidas, produced in the Bengali language, was released in 1932. Enormously popular in Bengal, it put Bose in the front rank among directors. Its showings were necessarily limited in other areas of India. But Bose followed it in 1933 with *Puran Bhagat* (The Devotee), in Hindi, which "put New Theatres on the all-India map and created a veritable sensation."[5] Such was the appeal of its music that it scored successes even in non-Hindi-speaking areas. In 1934 Bose's *Seeta* (Seeta), which had been made in Bengali the previous year, became the first Indian film to be shown at the Venice festival. In 1937 his *Vidyapathi* (Vidyapathi), another film about a Vaishnavite poet-saint, was released in both Bengali and Hindi versions, and scored new triumphs. A musicologist of the University of Delhi has put *Vidyapathi,* along with *Chandidas,* among the "revolutionary classics" which altered "the conception of the quality and function of music in a film."[6]

[4] A print of *Chandidas* has been preserved by George Eastman House, Rochester, N.Y. It may be mentioned that Nitin Bose, cameraman for *Chandidas,* later directed a Hindi version for New Theatres. He and his brother Mukul Bose, sound engineer for New Theatres, introduced into India the prerecording of songs, to liberate the camera during musical numbers. Nitin Bose later went to Bombay and became one of its leading directors. He is a first cousin once removed of Satyajit Ray.

[5] Bhanja and N. K. G., "From *Jamai Sashti* to *Pather Panchali,*" in *Indian Talkie, 1931–56,* p. 83.

[6] Bhatia, "Film Music," *Seminar,* December, 1961.

The impact of some of these films may be suggested by the pilgrimage of Kidar Sharma. As a student at Punjab University, in northwestern India, he had just finished his work for the M.A. in English when he saw *Puran Bhagat*. The film was such a "revelation" to him that he set off on a 1,500-mile journey from Amritsar to Calcutta solely in the hope of working with Debaki Bose. Arriving in Tollygunge, he found New Theatres a growing, bustling organization. Almost entirely self-sufficient, it already had a payroll of several hundred people. Sharma served briefly as an extra, then was able to hang on as sign painter.

One day, as he painted a billboard, B. N. Sircar parked his car beside Sharma's ladder. Getting out, he expressed approval of the work. Sharma said: "If you would take this brush from my hand, sir, and put a pen in its place, I might show you something better." B. N. Sircar, amused, gave him an adaptation exercise. All this eventually enabled Kidar Sharma to work under his guru, his mentor, Debaki Bose, on the production of *Vidyapathi*. Meanwhile it also won him a chance to write Hindi dialogue for what became another India-wide New Theatres sensation, *Devdas* (Devdas). But this brings us to a man who played perhaps the most spectacular role in the rising fortunes of New Theatres—the director of *Devdas*, Prince Barua.[7]

Pramathesh Chandra Barua, son of the Rajah of Gauripur, was born in 1903 in Gauripur, Assam. After graduating in 1924 from Presidency College, Calcutta, the handsome young Prince left on a European tour during which he took interest in all the arts including film, delighting especially in the works of René Clair and Ernst Lubitsch. Returning to India, he faced the problem of what to do.

He had innumerable interests, and everything came easily. An avid reader and music lover, he was also considered outstanding as horseman, marksman, dancer, tennis player, billiard player, hunter. In his native Assam he had already bagged several dozen tigers, a rhinoceros, and innumerable boars—although it is said he blanched at the sight of a cockroach.

[7] Kidar Sharma also became a leading Bombay director.

DIRECTED BY DEBAKI BOSE: *Vidyapathi* (VIDYAPATHI), WITH CHAYA DEVI AND KANAN DEVI, 1937. NEW THEATRES
COURTESY AURORA FILM CORPORATION

During the dyarchy period he served for a time, by appointment, in the legislative council of Assam. But the heady life of Calcutta was more to his liking. Settling there, he soon became involved in the film world. A newly formed producing organization, Indian Kinema Arts, persuaded him to play the villain Sarit in its silent film *Bhagyalaxmi* (The Blessings of Wealth).[8] He made a small investment in Ganguly's British Dominion Films and also acted a small part for him.[9] These experiences made film an incurable obsession. Returning to Europe, he got permission to observe production at the Elstree studios in London, then went to Paris, purchased lighting equipment, returned to Calcutta, formed Barua Pictures Ltd. and

[8] *Who Is Who in Indian Filmland*, pp. 23–25.
[9] Interview, Ganguly.

built a studio. It was Barua Pictures that produced *Aparadhi* (The Culprit), already mentioned as a critical success of the last days of the silent film. The first Calcutta production using artificial lights, it starred Barua and was directed for him by Debaki Bose.

When British Dominion Films collapsed, Barua also engaged Ganguly. To all appearances, Barua was about to rally around him various talents set adrift by the distintegrating film world of Calcutta. But Barua was actually no more ready for sound than British Dominion Films had been. His film activities had angered his father, the Rajah of Gauripur, who declined to help him. The Prince's ample living allowance, along with loans and investments by friends, had given him a start. But his plans and ambitions called for firmer footing. In the end Barua, like Ganguly and Bose, threw in his lot with New Theatres. From then on B. N. Sircar saw to it that the Prince had the budgets he needed.

Barua proceeded to justify Sircar's judgment in sensational fashion. The film *Devdas,* based on a popular Bengali novel by Sarat Chandra Chatterjee, adapted and directed by Barua, "revolutionized the entire outlook of Indian social pictures."[10] *Devdas* was first made in the Bengali language, with Barua in the lead; then it was remade in Hindi, again directed by Barua but with a recent New Theatres recruit, Kundanlal Saigal, playing and singing the role of Devdas. Both versions were released in 1935; ever since then, the Indian press has used the word "immortal" for Barua, Saigal, and *Devdas.* The songs as sung by Saigal became phonograph records that are still played on the radio systems of India, Pakistan, and Ceylon. Barua went on to produce numerous popular successes but was always afterward referred to as the man who had made *Devdas.* Even as director, Barua won the adulation reserved in later years for stars. In 1936 the film was remade by New Theatres in Tamil, again successfully. The cameraman for *Devdas,* a young Bengali called Bimal Roy, eventually went to Bombay, became one of its leading directors and producers, and in time produced a Hindi *Devdas* of his own, released in 1956. This, in turn, contributed to the rising fame of its stars, Dilip Kumar and Vyjay-

[10] Bhanja and N. K. G., "From *Jamai Sashti* to *Pather Panchali,*" in *Indian Talkie, 1931–56,* p. 83.

anthimala. Thus *Devdas* brought fame to many people. And virtually a generation wept over *Devdas.*

Although romantic-tragic in plot, the script for *Devdas* achieved a naturalness of tone that was, in its day, almost revolutionary. When Kidar Sharma completed the Hindi version of the Barua scenario, one reaction was: "This isn't dialogue, this is the way we talk."[11] It was precisely what Barua had wanted.

In Indian drama, such dialogue had never been an objective. Dramatic literature had long been associated with the language of courts. Perhaps for this reason, dramatists in the vernacular tended to write in a florid style, reaching for a remoteness associated with status. But Barua had been exposed to European naturalistic trends and wanted to put aside such language. He also demanded from his actors a quiet, natural tone. An actress who joined New Theatres from another company was astonished at how quietly the actors talked.[12]

It so happened that Saigal, the new discovery who played and sang the lead in the Hindi *Devdas,* had a sore throat when the songs were filmed. He had been a typewriter salesman, earning Rs. 80 per month, when Sircar, interested in his singing voice, offered him Rs. 200 to join New Theatres. But when he began to sing for *Devdas,* his voice cracked. The songs were postponed, but the sore throat persisted. Finally he tried the songs in a quiet, soft tone. It fitted the acting style Barua was trying to achieve, as well as the volume limitations of microphone and sound track. In this way, partly by accident, was born a singing style that soon spread over India, and that somewhat resembled a simultaneous Western development, the microphone crooner.

In the story of *Devdas* the hero—like Barua himself—was the son of a wealthy family of the ruling zamindar class.[13] In the story

[11] Interview, Kidar Sharma.

[12] Interview, Khote.

[13] The zamindar was an official instituted by the Muslims and retained by the British. He operated a tax-collecting concession in a specific territory. He paid the British-controlled government a stipulated annual revenue; collections from the people of his district were his concern. Zamindars were powerful, often very wealthy. Some became benevolent despots in their territories. The zamindar was often treated respectfully in films of the 1930s but since independence has become a favorite villain of period films.

Devdas falls in love with Parbati, with whom he has played since childhood, and who is the daughter of a poor neighboring family. The time comes for Devdas to go away to Calcutta for university studies. The parting is deeply felt by both. While he is away, her father arranges a marriage for her. She loves only Devdas but, obeying her father, prepares to suffer in silence the role of dutiful Hindu wife. The marriage takes place. Devdas, as a result, takes to drink. Among those who befriend him in Calcutta is Chandra, a "dancing girl" or prostitute, who is so anxious to save him that she is willing to give up her profession for him. Parbati, hearing of his decline, comes to see him to try to steer him away from his life of drinking. He says that, in the hour of final need, he will come to her for help. She returns to her life of duty. Eventually comes a day when a man, haggard from long illness, is found dead outside the high walls of her house. His remains are burned at the funeral ghats. Within her walls, Parbati hears the news that her Devdas is dead.

There are several things to be said about this story. Its "tragic" ending is at variance with Indian classical tradition, which permitted only a happy ending. Tragic endings were not used in Sanskrit drama and are even considered to be at odds with the Hindu view of existence. A life can hardly be interpreted as tragedy when life itself is only a transitional state. However, the tragic ending had become common in India before *Devdas,* especially in Bengali literature and drama, but this was apparently part of their European rather than Indian heritage.

To some extent *Devdas* was a film of social protest. It carried an implied indictment of arranged marriage and undoubtedly gave some satisfaction on this score to those who hate this institution. The powerful appeal of *Devdas* to the young must have been based in part on this element. Yet once this theme has set the story in motion, the film *Devdas* seems far less interested in the social problem than in the suffering. There was more than a little of German *Weltschmerzliebe* about *Devdas,* and it seems to have done much to popularize the doomed hero in Indian film. Doom itself has appeared to become irresistibly attractive. Very often artist and audience have needed only the flimsiest justification to believe in, and share, the doom of a hero. If Western films have erred toward forced

Synopsis of the story.
Grihadaha.

Characters.

Suresh – Young man – rich – loving nature, temperamental, kind hearted, conscientious, never does anything unless he is convinced, impulsive. Easily excitable if approached from weaknesses.

Has enough money – & lives comfortably though not luxuriously – never brags or swanks. Neat in appearance & habits.

FROM A BARUA NOTEBOOK: THE PRINCE WRITES A PART FOR HIMSELF

happy endings, some Indian films have favored forced unhappy endings.

Barua wrote most of the screen plays he directed. He left notebooks full of carefully penciled notes, plots, character sketches. Many seem to have been projections of his own concerns and conflicts. He was deeply concerned about the tragic dilemmas of his native land, its almost unbelievable extremes of wealth and poverty, of spirituality and cruelty. His plots often touched on these dilemmas but were, in the end, constructed to evade them.

An example was *Adhikar* (Authority), voted by the Film Journalists Association the best film of 1938.[14] Besides writing and directing it, Barua appeared as Nikhilesh. The story tells of Radha, a girl of the city slums, who longs for wealth and happiness. When she learns that she is the illegitimate daughter of a rich man, whose other daughter Indira, born of legal wedlock, is living in utmost luxury, she goes to Indira and claims half their father's estate. In-

[14] *Dipali,* January 27, 1939.

DIRECTED BY NITIN BOSE: *Bhagya Chakra* (WHEEL OF FATE), WITH UMA SASHI AND PAHARI SANYAL, 1935. NEW THEATRES

dira, shocked at what she learns, gives her shelter and money. Not content with this, Radha now brings about an estrangement between Indira and her fiancé Nikhilesh, and ultimately goads her half sister into giving her not only half but the whole of their father's estate. However, her unscrupulous ways turn the world against Radha and in the end she learns to her dismay that even her boy friend of the slums, Ratan, no longer cares for her.

By making Radha, in the later portions of the story, an increasingly unscrupulous character, the dramatist managed to remove the spotlight from the problem of social and economic disparity with which he confronted us at the start. In the end, he left matters weighted on the side of the status quo. Barua, aware of many issues but enamored of doom, was not essentially a radical force.

When he had finished a film, Barua could seldom endure to stay for its premiere. He would predict its utter failure, and then be off —for the forests of Assam, for Europe, or for America. In time he

DIRECTED BY P. C. BARUA: *Maya* (ILLUSION), WITH PAHARI SANYAL AS THE DOOMED HERO, 1936. NEW THEATRES

would be back, with notes for a new film. The films were almost always successful.

While some Barua productions brought into Indian film a note of sophisticated humor, his major successes were in romantic-tragic drama. Bombay financiers repeatedly offered him substantial sums to produce such films in Bombay. Pahari Sanyal, a star actor at New Theatres, was intermediary in one such offer. Barua waved it aside. He could not think of making films in Bombay, he said. "It is not my field. It is a bazaar."[15]

Admiration for him took extravagant tones. In an "open letter to Prince Barua" published in a Calcutta paper, a fan wrote:

> We are inclined to include you in the category of the great thinkers of the present day. By producing immortal *Devdas* you opened a new way for the Indian Film Industry, and since then you are looked on as a great philosopher.[16]

[15] Interview, Pahari Sanyal. [16] *Dipali*, January 27, 1939.

The tone of this letter was normal for such correspondence. In the same year, an observer wrote about the influence of stars on clothing fashions:

> Who can deny that Kanan's novel way of hair-dressing in *Mukti* has been "the method" of dressing for modern girls? . . . that Barua's curious cap in the same picture has won Calcutta-wide recognition as the most "up-to-date" headwear? . . . that Lila Desai's dancing sari in *Didi* is in vogue as "Lila sari"?[17]

In the film *Mukti* (Liberation), referred to here, Barua played the role of a romantic young artist who, to free his wife for another marriage, carries out a perfectly simulated suicide and then vanishes to the forest land of Assam. The scenes in Assam were shot on location there. When the wife and her new husband go to Assam in a hunting party, they all meet again. The artist rescues her from kidnapers and, in the course of this, is killed. Thus he has again given her *mukti*.

Barua himself married two wives—multiple marriages were still legal. The two wives lived in adjoining villas off Ballygunge Circular Road, Calcutta. Each bore him three children.

Barua was an unrelenting worker, but is said to have been always unruffled, considerate, debonair. He planned his work minutely. Unlike most Indian directors, he never *showed* an actor how he wanted a scene played. He had objections to making an actor a mimic, rather than an interpreter.

His influence as a director was immense, but its value is not easy to assess. It has been said of the films of his fellow director at New Theatres, Debaki Bose, that those films were of Bengal and could not have been made by any one other than a Bengali. Barua's films were not truly of Bengal, or of Assam, or of Europe, but hovered in a rootless world. Or so it seems, in retrospect, to many observers. But clearly his followers felt no such unreality, and readily shared his fantasies of doom.

Barua's best work was done during his first decade with New Theatres. In the 1940s he planned an ambitious Indian version of *The Way of All Flesh* but was not able to carry it out. He drank a good deal and people began to say that he had, at the start of his career,

[17] *Ibid.*, Anniversary Number, 1939.

P. C. BARUA: THE YOUNG PRINCE

P. C. BARUA AND KANAN DEVI IN *Mukti* (LIBERATION), 1937. NEW THEATRES

dramatized his autobiography. His health declined rapidly, and he underwent an operation in Switzerland. He returned full of plans, but soon collapsed. When he died in 1951, the *Journal* of the Bengal Motion Picture Association carried this obituary: "Pramathesh Chandra Barua, creator of *Devdas,* died at 4 P.M. on Thursday, November 29 last, at his Calcutta residence after protracted illness. He was 48."[18] Even in death, Barua was pulled back to his early triumphs, to the years of the rise and influence of New Theatres Ltd.

The self-made men of Prabhat

The second group that exerted immense influence on the Indian film in the early sound era was the Prabhat Film Company. Launched in 1929 in Kolhapur, it moved to Poona in 1933. Both cities are in the Marathi-speaking area.

If New Theatres was remarkable for the educational level of its leaders, the Kolhapur group was remarkable in an opposite way. Rich in talented directors, it had few with formal education.

One of its leaders, Rajaram Vanakudre Shantaram—known professionally as V. Shantaram—was born in Kolhapur in 1901. In his early teens he got a job in a railroad repair and maintenance workshop, where his working day was 8 A.M. to 6 P.M., and his salary Rs. 15 per month. At sixteen he acquired an additional job. Each day he went from the railroad shop to a local tin-shed cinema. Here, at a starting wage of Rs. 5 per month, he did odd jobs, eventually graduating to doorboy. He also painted signs.[19]

At the cinema, the stream of films was his education. Shantaram became saturated in the lore of the film world and closely studied film personalities. As a boy he was admired as a mimic of Western screen favorites: the French Zigomar, Max Linder, and Protea, and the Italian Foolshead. His Foolshead portrayal—his specialty—included not only the comedian's mannerisms but the quivering of the primitive screen image. The fascination of these creatures of the exotic Western world was, however, overshadowed by something else. Among the great events of his childhood was the

[18] *BMPA Journal,* December, 1951.

[19] Interview, Shantaram.

periodic arrival of a Phalke film. In the manner traditional to traveling shows, it would be promoted by a parade through the streets, with proclamations heralded by the beating of drums. It was not by accident that Shantaram's first sound film, years later, was *Ayodhyecha Raja* (The King of Ayodhya)—the story of Harischandra, the tale that had launched Phalke's Films in 1913.

The theatre job put the boy Shantaram in the midst of an exciting world, and led to a job as assistant to a photographer. Then, in 1921, he was hired by a new film company just starting in Kolhapur, the Maharashtra Film Company. Spurred by Phalke's success, it was producing mythologicals and historicals on regional history. Its proprietor, Baburao Painter, put Shantaram to every conceivable task in film production: cleaner, errand boy, scene painter, laboratory assistant, special effects man, camera assistant, performer. He emerged a superb and confident craftsman.

It was in 1929, at twenty-eight, that Shantaram with four partners decided to launch the Prabhat Film Company. His partners were V. G. Damle, K. R. Dhaiber, S. Fathelal, and S. B. Kulkarni. They began in a canvas studio, acquiring a tin-built studio only years later, after the move to Poona. For actors they at first leaned heavily on people of Kolhapur, and soon had a roster of a hundred local performers. "They accepted a rupee or two if we offered." The spirit of community participation was infectious. "When we needed elephants, horses and soldiers, the Maharajah of Kolhapur lent us as many as we needed. He even arranged mock battles and supervised the shooting."[20] But gradually the emphasis shifted to professionalism and the group transformed itself into a large, self-sufficient organization of several hundred artists, technicians, and assistants.

After a few silent films, the young company began in 1932 to release a stream of sound films in the Marathi language, some of which were also made in Hindi. Many later films were made only in Hindi. Some Tamil films were also made.

The three releases of 1932 included Shantaram's *Ayodhyecha Raja* (The King of Ayodhya), produced in both Marathi and Hindi versions. The role of Taramati, played in the Phalke film by the male actor Salunke, was this time played by a high-caste girl of Kolhapur,

[20] Fathelal, "Prabhat Was a Training School," in *Indian Talkie, 1931–56,* p. 139.

DIRECTED BY SHANTARAM: *Ayodhyecha Raja* (THE KING OF AYODHYA), WITH DURGA KHOTE, 1932. PRABHAT

Durga Khote, who was hailed as "the most spectacular newcomer of the year,"[21] and was to become one of India's most celebrated actresses.

Though studio facilities in Kolhapur were primitive, the company worked with precision and a sense of organization that in later years became unusual in the Indian film world. Durga Khote recalls:

> In those days the scripts were completely ready. . . . We knew, far in advance, what songs and what dialogue we were to render. . . . At Prabhat we were to report for work at 5:30 in the morning and we knew that shooting would be definitely over by 4:30 in the afternoon. There was no departure from this routine as shooting was done in sunlight and no artificial lights or arc lights were in vogue then. . . . We were called for work at 5:30 because it took two full hours for make-up with the hard grease of old days. By 8:00 we were ready for "take" and by tea-time we used to pack up as no more shooting was possible with the fading sunlight.[22]

[21] *Who Is Who in Indian Filmland,* p. 48.
[22] Khote, "Pictures Were Better Planned in Earlier Days," in *Indian Talkie, 1931–56,* p. 123.

Shantaram's first sound film won him wide recognition. From the start he struggled against the staginess of early sound films and tried to maintain the mobility of cinema. In 1936 his impressive spectacle, *Amar Jyoti* (Eternal Light), made in Hindi, was shown at the Venice film festival. But Shantaram's partners also shared in the rising glory of Prabhat. The following year *Sant Tukaram* (Sant Tukaram), which had been produced in Marathi in 1936, directed by V. Damle and S. Fathelal, became the first Indian film to win a Venice festival award.[23]

Sant Tukaram[24] tells with simplicity of a poet-saint of the village of Dehu, in the seventeenth century, who holds villagers spellbound with his songs of devotion—although his wife scolds him for impracticality and tells him he should be a better provider for their children. Tukaram is also beset by a jealous priest and would-be saint, Salomalo, who launches a variety of plots against him. Divine intervention rescues him—and the villagers—from several crises brought on by Salomalo. In due time great leaders come from afar to sit at Tukaram's feet; he resists offers of wealth for himself and his family. When the day of Nirvana comes for Tukaram, a heavenly vehicle arrives to take him. He invites his wife to go with him but she decides she is happy with her modest home, her children, and her buffalo, and elects to stay on earth.

The film has a charm and directness that set it apart from more pretentious devotional films of later years. Tukaram's conflict with his wife has an earthy believability and humor. He is no cardboard saint and she no mechanical shrew. Here the film medium derives strength from a sense of continuity with Marathi regional culture, just as the Debaki Bose films drew strength from Bengali culture. The film was an enormous success in the Marathi area, running fifty-seven weeks at one theatre in Bombay.[25] It had limited showings in other areas.

To Prabhat, as to New Theatres and other groups, well-known mythological and devotional stories seemed the safest starting point in the sound era. For decades an Indian producer, asked

[23] Holmes, *Orient,* p. 18.
[24] A print of *Sant Tukaram* has been preserved by George Eastman House, Rochester, N.Y.
[25] *Indian Cinematograph Year Book, 1938,* pp. 15–16.

DIRECTED BY SHANTARAM: *Amar Jyoti* (ETERNAL LIGHT), 1936, SHOWN AT VENICE FILM FESTIVAL. PRABHAT

why a film was popular, was likely to say, "Because the people know the story." Familiarity, not novelty, was long considered the safest investment.[26]

Even so Prabhat began to experiment with social films, and the most notable of these came from Shantaram. Never losing his taste for mythology, he also became known for tense stories on contemporary problems.

In *Duniya Na Mane* (The Unexpected), produced in Hindi in 1937—and also produced in Marathi under the title *Kunku*—we learn of a bright young girl, Nirmala, who is married to an elderly man through negotiations conducted by her uncle. In a series of deftly treated episodes, the old husband becomes aware of the enormity of the wrong done to her. Finally, in his eagerness to restore to Nirmala the freedom taken from her—divorce was not possible at this time—he kills himself. The film makes a good deal

[26] The first five years of sound brought three different versions of the Tukaram story, all in Marathi; and eight versions of the Harischandra story, in five Indian languages. *Indian Talkie, 1931–56,* list, pp. i–xxvii.

DIRECTED BY SHANTARAM: *Amar Jyoti,* WITH DURGA KHOTE, 1936.
PRABHAT

of use of symbolism, such as that of an old clock. The husband's "supreme act of atonement," as a reviewer called it,[27] leaves the girl —and the audience—with another problem to contemplate: the dread attached in Hindu society to widowhood. The film, concise and direct in statement, had wide impact and was repeated for years. It probably showed the influence on Shantaram of the resounding successes of Barua.

In 1939 Shantaram's *Admi* (Life Is for Living), produced in Hindi, again carried an implied challenge to traditional attitudes. It told of Moti, a police constable, who is assigned to raid a gambling den and brothel. Here he meets Kesar, a prostitute, and finds her eager to escape the vicious atmosphere of her life. He lends her a helping hand and eventually comes to love her. But the very religious atmosphere of his house makes the girl conscious of her shortcomings and drives her away. Moti is desperate and at last goes in pursuit of her.

[27] *Hindu,* May 12, 1939.

This film, too, was shown for years and won Shantaram a wide reputation. While it had more professional gloss than earlier Prabhat films, such as the simple *Sant Tukaram,* it is possible that something important was meanwhile being lost. The story, while taking issue with orthodoxy, does so through an artificial plot of surface attraction. The sentimentalized prostitute Kesar belonged more to the rootless world of Barua than to the Maharashtrian world in which Shantaram and his associates grew up. By necessity Shantaram, producing in Hindi, a language foreign to him, for a huge audience he did not know and whose entertainment requirements were made known to him via distributors, statistics, and trade press, was moving into a world of quasi-realistic fantasy. Shantaram was to remain for decades a figure of stature in the Indian film, often on the edge of world success. He will reenter our story many times, always commanding respect but not quite achieving the great goals before him. The reasons may already have been present in the first decade of Prabhat, when it was rising from regional obscurity to national attention and achieving, in a complex of huge studios on the edge of Poona, the status of large enterprise.

The road to Bombay Talkies

Late in 1933 two young people arrived in Bombay from London, bringing with them a completed film which had been finished in London and already shown there. Along with its English version they brought a Hindi version—not yet shown. The two young people were Indian, though both had been absent from India much of their lives. In their plans, much depended on the reception they would get in Bombay for their Hindi film. For both, long journeys had led to this moment.

Devika Rani Chaudhury—better known as Devika Rani—was born in Waltair in southern India. Her father, Colonel Chaudhury, soon afterward became Surgeon-General of Madras. A great-uncle on her mother's side was Rabindranath Tagore. When she was nine, her father shipped her to England to be educated, telling her she must learn to take care of herself. Attending South Hampstead School, and already showing rare beauty, she was befriended by an international set and discussed her career problem earnestly with various people. Should she be a dancer, as Anna Pavlova urged? A singer? A doctor, like her father? When she finished school she won a scholarship to the Royal Academy of Dramatic Art. She then became interested in architecture, enrolling in a course. She meanwhile earned surprising sums making Paisley designs for a British textile company, and wrote her father that he need not send her any more money since she could take care of herself. Then she met Himansu Rai, Indian film producer, and asked him what she should be.

Three fields, he said, would grow in importance, and in any of them she would be able to serve India: the press, radio, and film. He then offered her a job as consultant on costumes and sets for his next film, to be shot in India. The following year, after the shooting was finished there, they were married in southern India. She was still in her teens.

Himansu Rai was at this time in a crisis in his own career. He had been born in Bengal, where he was part of a large family that had its private theatre. He studied at the University of Calcutta, acquired a law degree, and also studied under Tagore at Santiniketan—the abode of peace, "where the world becomes one nest." During his time there, Gandhi visited the school. The young man was inclined to the arts, but the family's plans now called for him to go to London for training as a lawyer in the Inner Temple—as Gandhi had done. Himansu Rai gladly went.

In London he did his law work but could not keep away from theatre, and was soon carrying a spear in the fabulously successful musical *Chu Chin Chow*. Then he played a role in a London production of *The Goddess,* a play by a young Indian writer, Niranjan Pal. Meanwhile plans were forming.

He wanted to make a series of films on the great world religions. One would deal with the Buddha. Another would be based on the Oberammergau Passion Play. Rai journeyed to Munich, solely to promote this plan.

He was a skillful pleader and, early in 1924, persuaded the Emelka Film Company, of Munich, to take part in an ambitious project: an international co-production, the first plan of this sort to involve India. Telling the story of Buddha, it would be based on *The Light of Asia,* the poem by Edwin Arnold. Emelka would send to India a director, cameraman and assistants, and would provide all equipment. Its laboratory in Munich would process the film, and it would do all the editing. Emelka would own all European distribution rights. Himansu Rai pledged himself to provide an Indian cast and to raise funds in India to pay it and other location costs. The Indian investors would own Indian distribution rights and would receive from Emelka two prints to exploit those rights.

Himansu Rai went to India and, according to plan, raised an unusually large sum—eventually about Rs. 90,000—for production outlay. Work was begun, with Buddha played by Himansu Rai and the feminine lead by a thirteen-year-old Anglo-Indian girl, Sita Devi—actual name, Renee Smith. The scenario was by Niranjan Pal. The director, from Germany, was Franz Osten; the producer, Himansu Rai. The film had gala openings in Berlin, Vienna, Buda-

The Light of Asia, WITH HIMANSU RAI AND SITA DEVI, 1925.
HIMANSU RAI PRODUCTION FOR EMELKA

A Throw of Dice, WITH HIMANSU RAI AND SITA DEVI, 1929.
HIMANSU RAI PRODUCTION FOR UFA

pest, Venice, Genoa, Brussels, with personal appearances by Himansu Rai and toasts by international notables.[28] Throughout Central Europe the film was a financial triumph for Emelka. In London it had a Royal Command Performance and ran over four months at a concert hall, though not profitably.[29] In India the film received favorable criticism but only limited success at the box office. The trade persisted in considering it "foreign." The backers, moving their two prints from city to city, experienced mounting disillusionment. After two years, Rs. 50,000 remained unrecovered.[30] What Rai had hoped would launch a series of international productions actually closed the door to Indian capital.

But what had closed doors in India opened them wider in Ger-

[28] Interview, Madhu Bose, and *Evidence,* III, 1001.
[29] *Evidence,* II, 203.
[30] *Ibid.,* III, 1015.

many. The continental success of *The Light of Asia*[31] brought Himansu Rai offers that resulted in two more Indo-German productions, both shot in India, both performed by Indian casts including Himansu Rai and Sita Devi, both written by Niranjan Pal, directed by Franz Osten, and produced by Himansu Rai—this time with German capital. The first was *Shiraz,* made in 1926 under the Emelka banner, and telling a story of the designer of the Taj Mahal.[32] The second was *A Throw of Dice,* produced under the banner of Ufa—Universum Film Aktiengesellschaft, the giant government-subsidized organization with studios at Neubabelsberg. It was this film that involved Devika Rani.

When the shooting was completed for *A Throw of Dice,* Himansu Rai and his bride hastened to Neubabelsberg, Germany, for the editing. The world of Ufa was thrown open to them. As the editing progressed, Devika Rani became a trainee in the Erich Pommer unit. She got advice from Fritz Lang, watched the shooting of *The Blue Angel,* held the make-up tray for Marlene Dietrich, and engaged in intensive seminars with the great director G. W. Pabst. Pabst, in one exercise, would place a trainee before the camera and, while the action was shot, ask a series of questions to be answered with "yes" or "no." Then Pabst would review the film in minute detail. "Never use that one," he would say. "See what an ugly thing it does to your mouth there? Remember that, eliminate it."[33]

The Ufa training created exciting vistas. But one day Ufa issued a barrage of dismissal notices. The rushing advance of sound had brought need for a complete Ufa reorganization. Indian co-production no longer seemed feasible; the German career of Himansu Rai was ended. Not long afterward Ufa came under Nazi control. Lang and others were ousted in the interest of racial purity. Pommer and Pabst went into exile. Meanwhile Emelka of Munich, which had given Ufa some competition, became a victim of the spreading financial depression. The doors in Germany were closing.[34]

However, both *Shiraz* and *A Throw of Dice* won sufficiently favor-

[31] A print of *The Light of Asia* has been preserved in the National Film Archive of the British Film Institute, London.

[32] A print of *Shiraz* has been preserved by George Eastman House, Rochester, N.Y.

[33] Interview, Devika Rani Roerich.

[34] See Bardèche and Brasillach, *History of Motion Pictures,* p. 350.

able reception in England to open doors there[35] for Rai. Once more he launched an international production, this time Anglo-Indian, with English capital, and with sound. This was the beginning of *Karma* (Fate), in which Devika Rani co-starred with Himansu Rai.[36]

In 1930 the couple went to India for many months of exterior shooting and intensive study of Hindi; then they returned to London's Stoll Studios for the interiors, including the recording of the songs. For every shot, two takes were made: one in English, one in Hindi. Because of a limited budget, two takes were usually the limit, except for the songs.

The film took over two years to complete. It was a modern story about a beautiful young maharani (Devika Rani) who wanted "progress"—never explained—and her love for a prince of a neighboring Indian state (Himansu Rai) who also wanted "progress" but whose father, the maharaja, did not. Marriage brings her, by the rules of Hindu society, under her father-in-law's authority, and this creates the conflict of the film. It was premiered in London in May, 1933. Critics had some reservations about the story, which a few considered naïve, but all fell at the feet of Devika Rani. "A glorious creature," the *Era* called her. "Devika Rani's large velvety eyes can express every emotion."[37] The *News Chronicle* declared that "she totally eclipses the ordinary film star. All her gestures speak, and she is grace personified."[38] The *Star* reported that "her English is perfection."[39]

Fox Film Corporation now wanted Devika Rani to star in a film about Bali, and a German producer wanted her for a film about a snake charmer. But Himansu Rai said: "Let us learn from these people, but let us put the knowledge to work in our country."[40]

[35] Rotha, *The Film Till Now*, p. 325, found the films "singularly uninteresting." But British capital showed increasing confidence in Himansu Rai. A British distributor had guaranteed the German backers £7,500 for British rights in *Shiraz*. *Evidence*, III, 1004. British capital also gave Ufa an advance guarantee on *A Throw of Dice*, and was ready to take the full risk on Rai's next venture.

[36] A print of *Karma* has been preserved in the National Film Archive of the British Film Institute, London. Sita Devi did not appear in *Karma*. Her successes in three Rai films won her a starring position in the Madan company. Not having mastered Hindi, she slipped from the public eye after sound.

[37] *Era*, May 17, 1933. [38] *News Chronicle*, May 11, 1933.

[39] *Star*, May 15, 1933. [40] Interview, Devika Rani Roerich.

DEVIKA RANI IN *Karma* (FATE), 1933

Their film future, they knew, must be in India. With the advent of sound, this was more certain than ever. Such was the stake when *Karma,* in its Hindi version, had its Bombay premiere on January 27, 1934. Its reception once more opened the doors of Indian investors.

That year Bombay Talkies Ltd. was formed and a studio built. Under the painstaking supervision of Himansu Rai, it purchased the most modern equipment. In 1935 a stream of Hindi productions began to emerge from Bombay Talkies Ltd. Franz Osten, director of *The Light of Asia, Shiraz,* and *A Throw of Dice,* had joined the staff. A handful of other technicians came from Germany and England. Otherwise, the staff of more than four hundred artists, technicians, assistants, and others was Indian. It became, like New Theatres and Prabhat, a largely self-sufficient organization.

Mindful of the exhilarating days with the Pommer unit, Himansu Rai and Devika Rani soon instituted a trainee program. Each year Rai interviewed scores of job candidates, many sent by Indian universities. Within a few years the names of a number of younger Bombay Talkies staff members were known throughout India. They included actors Ashok Kumar (he began as laboratory assistant), Raj Kapoor (he began as clapper boy), Dilip Kumar; producer S. Mukherjee; writer K. A. Abbas.

Himansu Rai's desperate efforts for international co-production had perhaps been ahead of their time. For years to come, with the difficulties of sound, film producers in most countries would concentrate on home problems. Bombay Talkies Ltd. would do likewise. But co-production would, in later decades, once more emerge as a challenging and necessary idea, and in India the pioneer work of Himansu Rai would remain a reference point for all such ventures.

Bombay Talkies now settled down to a schedule of about three features a year. Some, like *Savitri* (Savitri), produced in Hindi in 1937, were mythologicals. This story from the *Mahabharata* had already been the basis of five different sound films in four Indian languages,[41] but the Bombay Talkies version was admired for its

[41] *Indian Talkie, 1931–56,* list, pp. i–xxvii.

unique delicacy. The story tells of Savitri, daughter of King Ashwapati, who ruled thousands of years ago. Savitri marries a man fated to die. But she so much loves her husband that by her piety, penance, and devotion she wins back his life from Yama, the god of death. The story has a quality akin to that of the Orpheus story—but with a happy ending.

Other Bombay Talkies films were socials, like *Achhut Kanya* (Untouchable Girl), made in Hindi in 1936. The highly respected newspaper the *Hindu,* which had begun a weekly page of film news and comment, considered it "easily the best Devika Rani film to date," and described it as follows:

> She appears as a Harijan girl, in love with a Brahmin youth, portrayed by Ashok Kumar. Caste barriers and religious bigotry stand in the way of their union. The boy is forced into a marriage to a wife he cannot love and the girl to one of her own class. Wisely, afraid of their love, they keep out of each other's way, till chance throws them together at a village fair. Inflamed by jealousy and egged on by neighbours, the girl's husband mistakes this meeting and a fierce encounter ensues between the Brahmin youth and the untouchable husband on a level-crossing. A train comes on. The girl, in an effort to part the combatants, is run over and killed, a human sacrifice at the altar of bigotry.[42]

Here, in most striking form, are all the most representative features of the Indian social film of the first decade of sound. Its writers were perhaps too apt to grasp at the neat plot, the ready-made ironies, the popular yearning for doom. On the other hand, they did not shrink from stories that attacked, implicitly and often explicitly, the canons of Hindu society. This was a time when law and precedent obstructed the intercaste marriage and firmly supported the ostracism of the untouchable. To be sure, Gandhi, Nehru, and other leaders of the Indian National Congress called for an end to this and repeated, week after week, that independence in itself would not be enough, that Hindu society must also reform itself from within. They ceaselessly hammered at issues of caste, untouchability, widow remarriage. On the level of popular fable, the social films did likewise.

What was their impact? Neither historian nor sociologist can give us the precise answer. P. R. Ramachandra Rao, reviewing the

[42] *Hindu,* June 25, 1937.

Achhut Kanya (UNTOUCHABLE GIRL), WITH DEVIKA RANI AND ASHOK KUMAR, 1936. HIMANSU RAI PRODUCTION FOR BOMBAY TALKIES

achievements of the Indian film at a twenty-fifth anniversary celebration of the industry, nevertheless felt that he knew. Praising the Indian film for "rapid strides," he declared: "It has unsettled the placid contentment of the Indian masses, it has filled the minds of youth with new longings and it is today a potent force in national life."[43]

[43] *Ibid.*, May 19, 1939.

In few places of the world would a film industry have been praised for unsettling public contentment. Yet it is possible that film was indeed shaping attitudes, and doing so in a variety of ways. It was known that at Bombay Talkies all company members, of whatever caste, ate together at the company canteen. It was even said that top actors, on occasion, helped to clean floors. Himansu Rai, perhaps with memories of Gandhi's visit to Santiniketan, insisted on such scrambling of functions. All this was part of the legend and role of Bombay Talkies.

Years later the government of independent India was to bestow on Devika Rani the title of Padma Shri, and the magazine *Filmfare* took the occasion to look back to the 1930s, to the first days of Bombay Talkies, declaring: "It was but natural that Bombay Talkies soon came to stand for new values."[44]

By the time those words were written Bombay Talkies Ltd., like its contemporaries New Theatres and Prabhat, had long ceased to exist. They and others had been swept away by new and powerful forces. The nature of those forces must be examined later.

The changing film map

We have mentioned New Theatres, Prabhat, and Bombay Talkies as pace-setters and molders of taste in the first decade of sound. There were a number of other companies, of similar structure, that achieved successes in this period. We shall list the most influential.

It should be emphasized that the geographical pattern of the Indian film industry was, throughout the decade, undergoing changes. It is important to understand this development.

Almost continuously two opposing trends were at work. One tendency was for each language area to develop a production center or centers of its own. This trend appealed to regional pride, made efficient use of talent of acceptable accent, and built regional stars. It had the disadvantage that a wide range of technical services could hardly be supported by every language area.

For this reason there was an opposite tendency, resulting in a concentration in three large centers, each producing in the language

[44]Malik, "Padma Shri Devika Rani," *Filmfare,* May 14, 1958.

of its area but also attempting, often through imported talent, to reach into and exploit other language areas. A listing, by location, of the production companies of 1937 shows the two tendencies at work:[45]

Bangalore	2	Kolhapur	6	Poona	4
Bezwada	2	Kumbakonam	1	Rajahmundry	2
Bombay	34	Lahore	4	Salem	6
Calcutta	19	Lucknow	1	Tanjore	1
Coimbatore	8	Madras	36	Trichinopoly	2
Dharwar	1	Madurai	7	Tirupur	2
Erode	2	Nellore	1	Vizagapatam	1

The smaller centers usually began by concentrating on one language. The larger centers felt from the start that they must work also in others. Sometimes a film script, after proving a success in one language, would be reenacted in another, with an entirely new cast. In this way *Devdas,* after its success in Bengali, was repeated by New Theatres in Hindi and later in Tamil.

But a producer could save various costs by shooting two or more versions simultaneously. The "double versions" began almost immediately. In these, each separate shot is done first in one language, then in another. The operation may call for two complete casts, although dance numbers may serve both versions. The shout of "Bengali take!," followed a few minutes later by "Hindi take!," became common in Calcutta studios, while other combinations were heard in Bombay. Occasionally a bilingual actor might appear in both films of a double version. But the prevailing tendency was to use double casts, and this is one reason why film companies grew rapidly in size. The large companies acquired acting staffs representing two or more major languages.

Calcutta almost at once achieved a monopoly over Bengali production, using this as a base for forays into other language markets, especially Hindi. The Bengali-Hindi double version became a standard activity at New Theatres and other Calcutta companies—such as the East India Film Company, launched in 1932.

Bombay and nearby cities, including Poona and Kolhapur, meanwhile took charge of Marathi production, using this as a base for

[45] *Indian Cinematograph Year Book, 1938,* pp. 283–87. Lahore later became part of Pakistan.

incursions into Hindi and other language areas. Since Bombay was close to the Hindi area and had a fairly large sprinkling of Hindi-speaking people—such as factory workers who had migrated from rural areas—it was in a good position to take a prominent role in Hindi production.

The two leading languages of southern India, Tamil and Telugu, were for some years the focus of mighty struggles. When sound began in Bombay and Calcutta, there was no sound-production equipment or studio in Madras, the center of the Tamil area. The large Tamil market looked open to others. In 1932 and 1933, Tamil films were produced in Bombay by the Imperial Film Company, producer of *Alam Ara,* and a new company called Sagar Movietone; in Calcutta by New Theatres and the East India Film Company; and in Poona by Prabhat. For these films the companies usually arranged junkets of Tamil-speaking actors from Madras. This sort of activity stirred southerners into action.

Among those in Madras who had some film experience was K. Subrahmanyam, a young criminal lawyer with a passion for the arts. In the late 1920s, while getting a foothold in law, he had made side earnings selling stories for silent films to a newly formed company, Associated Films. This had been started by a professional strong man, Raja Sandow, who after playing hero roles for Chandulal Shah in Bombay decided to take up production in his native province. Associated Films soon "failed for want of business-like instinct,"[46] but meanwhile the young criminal lawyer had won local notoriety as a film expert. In 1934 a Madras financier, intent on producing a Tamil-language film, invited Subrahmanyam to write and direct it.

There were still no available facilities, other than three glass-roofed studios, and it was decided to shoot in the open air. One difficulty was that the financier had had a quarrel with a former business associate, who came each day and parked his baby Austin close to the production. As soon as he heard "Silence, please! . . . sound . . . camera!" he would start honking his horn. This persuaded the financier to settle with his former associate. And, although *Pavalakkodi* (Pavalakkodi) was completed and was a box-office success—it

[46] *Who Is Who in Indian Filmland,* p. 11.

had fifty songs—it persuaded Subrahmanyam that there were better ways of making films.

That same year an entrepreneur in Salem, T. R. Sundaram, took a group of actors to Calcutta, rented the Madan studio in Tollygunge for three months at a cost of Rs. 25,000, and completed a Tamil-language film that proved so profitable in the south that he went north for six more junkets, all profitable.[47] Meanwhile Subrahmanyam was offered financial backing for similar junkets, for which he rented the East India Film Company studio in Calcutta. On the first such junket he took sixty-five people, renting a three-story house for them for three months and a car to shuttle them to and from the East India studio. The studio supplied all technical personnel, including its editor.[48] All the films were financial triumphs. Other producers arranged similar trips to Bombay.

The vistas of expanding profits meanwhile spurred construction of up-to-date southern studios. Several such studios were built during 1935–36 in Madras, Salem, and Coimbatore. These included a studio built jointly by several Madras producers, organized as the Motion Picture Producers Combine. Thereafter Madras was never dependent on northern studios. The producers in the Madras area now began to take charge of Tamil production, and gradually also took control of production for the nearby Telugu area, as well as the important Kannada and Malayalam language groups. By the 1940s Madras, grown powerful through its grip on these markets, also began to make astonishingly successful forays into Hindi production. In the 1950s it would in some years pass Bombay in volume of production.

Thus Bombay, Calcutta, and Madras became the three major centers. Each had its own language specialties, but for each the Hindi market remained a target. Here lay the big stakes.

There is an irony and a problem in all this. As the successful 1930s drew to a close, Prince Barua, whose native tongue was Assamese and who also spoke fluent Bengali and English, was being pressed to make his supreme efforts in Hindi. Similarly in Poona, V. Shantaram, whose native tongue was Marathi, was necessarily doing

[47] Interview, T. R. Sundaram.
[48] Interview, Subrahmanyam.

his principal work in Hindi. The partners of Bombay Talkies, Himansu Rai and Devika Rani, both products of Bengali culture, were likewise concentrating on Hindi. Even in Madras, speakers of Tamil were becoming producers of Hindi films.

What was this Hindi?

Bengal prided itself on a long tradition of Bengali culture, which had already thrived four centuries earlier in the era of Chandidas and Vidyapathi. The Marathi language also had a long literary heritage, including revered poet-saints of the thirteenth century and the beloved Tukaram of the seventeenth century. As for Tamil, it claimed a literature going back at least to the fourth century.[49] But Hindi was a new development, of meager literary background. For many producers it was as devoid of associations as Esperanto. Yet the troublesome structure of the Indian language map demanded concentration on Hindi. If many observers have found in the Indian film an increasing rootlessness, an increasing divorce from reality, one reason may be that many of its finest talents have had to exert themselves in a language not their own, spoken by people from whom they were both physically and culturally removed. This became, and will remain, one of the agonies of the Indian film.

We now list some of the companies that achieved success in the decade of the 1930s. Although seldom reaching the quality of New Theatres, Prabhat, and Bombay Talkies, the following contributed to the growth of the industry and the shaping of trends.

In the Bombay area:

IMPERIAL FILM COMPANY. Already mentioned as producer of the first talking feature, *Alam Ara*. An aggressive company of varied output, it averaged seven features a year during the 1930s. It made films in Hindi, Marathi, Tamil, and Telugu; it made the first film in Burmese; for export to Iran, it made three films in Persian.[50] During the decade its staff numbered several hundred artists, technicians, and others.

An interesting Imperial actress, star of *Alam Ara,* was Zubeida, a Muslim princess. Her career was symptomatic of a fairly steady shift in Muslim attitudes toward film and drama throughout much

[49] Sastri, *History of South India,* pp. 355–59.
[50] *Indian Talkie, 1931–56,* p. 24.

of India. Zubeida, daughter of His Highness the Nawab of Sachien and the Begum Fatima, had been permitted to enter films at the age of twelve; she was nineteen when she starred in *Alam Ara*. "She looks very innocent and charming with her beautiful oblong face." Her two sisters also began screen careers in their early teens and eventually their mother, the Begum, turned to film direction and became "India's first lady director."[51]

A more important Imperial star of the 1930s was Sulochana, born Ruby Meyers. Earning Rs. 2,500 per month in 1933, she was reportedly the highest-paid actress in India.[52] A Sulochana smash hit of 1934, produced in Hindi, was *Indira, M.A.* (Indira, M.A.). Its central character was a highly Westernized Indian girl, complete with Master of Arts degree, who forsakes the fine young man she was engaged to marry and weds a Westernized wastrel, with unhappy results. Westernized characters in Indian films, wearing Western clothes, smoking, drinking, and interspersing their speech with English phrases—"Correct!," "If you please," "My dear fellow," "Don't mention it"—were usually foolish, villainous, or ridiculous.

SAGAR MOVIETONE. Already briefly mentioned. The proprietor, Chimanlal Desai, had begun as a retail coaldealer in Bangalore, then turned film exhibitor, branched into distribution, and eventually came to Bombay to form Sagar Movietone, about the time Irani was planning *Alam Ara*. Desai, as distributor, handled some of the *Alam Ara* road tours and was astounded at the public response.

Having theatre interests in southern India, Desai quite naturally turned to production in Tamil and Telugu. After the rise of production in Madras, Sagar Movietone concentrated on northern tongues, mainly Hindi. It produced the first film in Gujarati, and also produced in Punjabi. During the 1930s it averaged six productions a year. Never pioneering in theme or treatment, it scored substantial successes.

A Sagar success of 1937, made in Hindi, was *Jagirdar* (Jagirdar). As described in the film page of the *Hindu*, it had that free and unconcerned dependence on accident and misunderstanding that was—and still is—characteristic of the work of the less distinguished

[51] *Who Is Who in Indian Filmland*, pp. 49, 68, 72.
[52] *Ibid.*, p. 67.

SULOCHINA IN *Indira, M. A.*
COURTESY NATIONAL FILM ARCHIVE OF INDIA

producers. It told of a young man, Jagirdar, who secretly marries Neela, a village girl, but parts from her when he is "suddenly" called abroad. She receives news that he has been shipwrecked. She "becomes a disgraced mother" and attempts suicide, but is saved by a village boy who marries her and acknowledges the child as his own. Neela is very happy with her child, when her long-lost husband, having survived the shipwreck, returns. The *Hindu* reviewer adds: "This cleverly woven plot finds an end in a solution unexpectedly provided by destiny."[53]

WADIA MOVIETONE. Specialists in stunt films and also producers of mythologicals. Stunt films stemmed from the world-wide successes of Douglas Fairbanks and the serials of Pearl White, Eddie Polo,[54] and others. Stunt films continued to hold their popularity throughout the 1930s. They were generally period dramas full of struggles on the edges of precipices, replete with heroic action.

Typical of the genre was *Hunterwali* (Girl Hunter), a Wadia box-office triumph of 1935, made in Hindi. It tells of a princess who sets out to rescue her father, held captive by a scheming minister. She disguises herself as a man and roams the countryside, robbing the rich to feed the poor. She meanwhile meets a peasant boy and they fall in love. They fight side by side against overwhelming odds and eventually rescue the king.

Stories of this vein still command a vast audience in several language areas. In the 1950s we shall find such films playing a surprising political role.

Wadia Movietone was founded in 1933 by Jamshed B. H. Wadia, whose education included an M.A. in English and a law degree. He was a tutor in English at St. Xavier's College in Bombay before taking up heroic films. His brother Homi Wadia joined him, directing *Hunterwali* and many other Wadia successes.

Although their education seldom intruded on their stunt films, the Wadias pursued a notable side interest. Like a number of

[53] *Hindu,* August 20, 1937.

[54] The American Eddie Polo, now almost forgotten in the United States, was among the most popular film heroes in India in the early 1920s. In 1927 a headmaster of a high school in Hyderabad, Sind, told the Indian Cinematograph Committee: "I once asked my class of 50 boys what was their ambition in life. Five boys wrote, 'To be Eddie Polo.'" *Evidence,* I, 675. Many Indian producers today have vivid memories of the films of Eddie Polo.

other producers, they made occasional "topicals"—newsreel items—and other short films. But such material was becoming increasingly difficult to market. The Wadias, for a time, provided a newsreel free of charge to every exhibitor booking a Wadia feature. All this paved the way for the prominent role J. B. H. Wadia was to play in the Second World War in the field of government-sponsored documentaries and newsreels.

During the 1930s the Wadias produced in Hindi, Tamil, and Telugu, averaging three or four films a year.

MINERVA MOVIETONE. Founded by Sohrab Modi, an actor who had toured for years in a Hindi (Urdu) version of *Hamlet,* which in 1935 became a sound film. Modi, after studies at the University of Bombay, had entered the film world in 1914, handcranking a projector in a small cinema in Gwalior. He left to become a stage actor, but after the coming of sound returned to film as actor, director, and producer.

Minerva Movietone specialized in historic spectacles. Its greatest success was *Sikander* (Alexander), the story of Alexander the Great's campaign to the border of India, where he had to turn back. Made in Hindi in 1940 and released the following year, *Sikander* was approved by the Bombay censor board but was later uncertified for some of the theatres serving army cantonments. The mutiny of Alexander's soldiers was apparently not suitable entertainment for Empire troops. While *Sikander* had rousing battles and was rated by a British writer "well up to the standard of that old masterpiece *The Birth of a Nation*,"[55] its portrait of history was a shallow one of posturing heroes. However, its appeal to nationalism was so direct that it remained popular for years. It was revived in Delhi in 1961 during the Indian march into Goa.

RANJIT MOVIETONE. The production company formed by Chandulal Shah and "glorious Gohar" after their early successes with *Why Husbands Go Astray* and other silent films. Ranjit eventually acquired four sound stages and achieved an output of six features a year, which stretched over more than a decade of uninterrupted successes. The company concentrated on socials. Chandulal Shah took the lead in adopting Western promotion techniques, includ-

[55] Nichols, *Verdict on India,* p. 114.

NOT FOR BRITISH TROOPS. *Sikander* (ALEXANDER), 1941, THE RETREAT OF ALEXANDER THE GREAT FROM INDIA. MINERVA (HINDI)

ing mammoth posters and electric signs. He once sent three thousand telegrams to promote a film: "COME MEET MISS GOHAR IN 'BARRISTER'S WIFE.'" Shah, who began his working life in the Bombay Stock Exchange, continued to speculate and at one time tried to corner the cotton market—unsuccessfully.

During the 1930s Ranjit maintained a payroll of about 300 artists, technicians, and others. It produced in Hindi, Punjabi, and Gujarati.

Others in the Bombay area in the 1930s: BHARAT MOVIETONE, producing in Hindi, Marathi, and, until the rise of southern production, Telugu; SARASWATI CINETONE, producing in Hindi, Marathi, and, for a time, Telugu; KRISHNATONE, producing in Hindi and, for a time, Telugu; SHARDA FILM COMPANY, producing in Hindi.[56]

[56] *Indian Talkie, 1931–56,* list, pp. i–xxvii.

In the Calcutta area:

EAST INDIA FILM COMPANY. Founded in 1932 by R. L. Khemka, a dealer in automobile parts, who had an appreciation of good equipment. Its activities in the 1930s were varied and venturesome. Besides producing in Bengali, Hindi, Tamil, and Telugu, it made a film in Persian by importing a cast from Iran and another for the Malayan market by importing a cast from Singapore.[57] In the 1930s it had a payroll of 300 artists, technicians, and others, and an output averaging eight films a year.

Its early films in Tamil and Telugu involved the import of talent from southern India. This led southern producers to rent the East India Film Company studio for production junkets. When this in turn led to studio building in Madras, a number of East India Film Company technicians migrated south to the new studios. Calcutta technicians from East India Film Company thus played a prominent role in the rise of Madras as a production center.

Early in the Second World War the studios of East India Film Company were requisitioned for British army use.

Others in the Calcutta area in the 1930s: AURORA, a company that had begun as a traveling cinema in 1913,[58] operating in Bengal and Assam, and had later gone into production on a modest scale; during the 1930s it produced in Bengali, Assamese, Tamil, and Telugu; RADHA FILMS, producing in Bengali, Oriya, and, for a time, Tamil and Telugu; KALI, producing in Bengali, Oriya, and Telugu.[59]

In the Madras area:

MADRAS UNITED ARTISTES CORPORATION. Formed by K. Subrahmanyam in 1936 as a by-product of his production trips to Calcutta. His partner in the venture was S. D. Subbalakshmi, leading feminine star in those tours. During the late 1930s the company had a payroll of 350 artists, technicians, and others. This included an orchestra of 22 musicians which did radio concerts. During the 1930s the company produced in Tamil and Telugu, later expanding into Malayalam and Kannada.

The earliest Tamil and Telugu films were mythologicals, but Madras United Artistes soon added socials. Subrahmanyam, one of

[57] Interview, Khemka, and *Indian Talkie, 1931–56,* list, pp. i–xxvii.

[58] *Evidence,* II, 666.

[59] *Indian Talkie, 1931–56,* list, pp. i–xxvii.

TAMIL PRODUCTION: SUBRAHMANYAM'S *Balayogini* (CHILD SAINT), 1936, FEATURING A BRAHMIN WIDOW. MADRAS UNITED ARTISTES

the few Brahmins among early Madras producers, outraged the Brahmin community by the production of *Balayogini* (Child Saint), made in the Tamil language in 1936. It told of a Brahmin widow who, driven out by a rich relative, decides to live with a low-caste servant who shelters her and her little daughter. The story offended orthodoxy on several grounds, but Subrahmanyam compounded the offense by persuading a Brahmin widow to play the role of the Brahmin widow. Brahmin widows were expected to shave their heads, wear only white saris—always covering the head—and live a prescribed life of austerity and seclusion. The sight of a widow was a bad omen. The sight of a widow on the screen was defiance of taboo on a grand scale. The film was especially successful because it introduced a new child actress, letting her speak simple wisdom and give voice to a skepticism that, in effect, heaped ridicule on caste restrictions. The child actress at once became one of the most celebrated film personalities of southern India, and "perhaps there was[60]

[60] Gopalakrishnan, "Four Decades of Tamil Films," *Filmfare*, April 20, 1962.

not a single parent in the south who did not fondly hope that his little girl would blossom into another Baby Saroja."

A number of early Tamil films had an anticaste attitude, which helped them to win a wide popular following. It also, in conservative Madras, won for the film industry strong antipathy among those of Brahmin caste. This alignment played a role in the later film history of Madras and southern India.

MODERN THEATRES LTD. Founded in 1936 in Salem, 300 miles south of Madras. Its founder, T. R. Sundaram, had studied textile dyeing at Leeds University in England. But the rousing reception given to early Tamil films led him to plan the Calcutta production junkets already mentioned, using the Madan studio. The profit from those ventures—they included a *Krishna Leela* (Drama of Krishna) that had 62 songs and 3 dances—led to the building of a studio in Salem where Modern Theatres Ltd., with a staff of 250, settled to a long career of steady production, averaging three films a year during the

TAMIL PRODUCTION: SUBRAHMANYAM'S *Thyagabhoomi* (LAND OF SACRIFICE), 1938. BRAHMINS DRIVE HARIJANS FROM A TEMPLE. MADRAS UNITED ARTISTES

1930s. The company started in Tamil and Malayalam, later expanding into Telugu and Kannada.

VAUHINI PICTURES. Started in Madras in 1939 by B. N. Reddi, along with two associates who had been at the Prabhat Film Company. B. N. Reddi had a degree in accounting from the College of Commerce, Madras. But he had previously studied under Tagore at Santiniketan and been exposed to the Bengali theatre and film world. Reddi began his film career by specializing in Telugu. Much of his early work is said to have reflected the culture of Andhra, the Telugu-speaking area, in the way that the work of Debaki Bose reflected Bengal. Vauhini Pictures was to grow rapidly and to develop one of the finest groups of studios in Asia, maintaining high technical excellence.

Others in the Madras area in the 1930s: TAMILNAD TALKIES, start-

TELUGU PRODUCTION: B. N. REDDI'S *Vande Matharam* (HONOR THY MOTHER), 1939. VAUHINI

ing in Tamil but later adding other south Indian languages; SRINIVAS CINETONE, producing in Tamil; VEL PICTURES, producing in Tamil and Telugu; ROYAL TALKIE, producing in Tamil and Telugu.[61]

One big family

The Indian film world, at the end of the 1930s, was marked by a feeling of confidence. It was also beginning to have the look of an organized industry. A Motion Picture Society of India had been formed in 1935, followed rapidly by groups with regional emphasis: in Calcutta, the Bengal Motion Picture Association (1936); in Bombay, the Indian Motion Picture Producers Association (1937); in Madras, the South Indian Film Chamber of Commerce (1938). Each began to issue journals, bulletins, statistical volumes. The industry was by now the focus of a trade press of sixty-eight periodicals, of which half were in English, half in various Indian languages.[62] All in all, as the industry in 1938 took note of its twenty-fifth anniversary, it had a look of solidity and self-respect.[63]

The developing feeling of strength was furthered by the structure of the companies that, up to this time, dominated Indian film production. The big companies of the 1930s, like the Phalke company before them, seemed to be extensions of the joint-family system. Many of the companies had, in fact, clusters of relatives. In India this is not considered nepotism but normal, commendable family loyalty.

Each company had a wide range of personnel and almost never had to turn to outsiders for help or services. Each had its own laboratory; this was long assumed to be essential. Each had its studio or studios and its preview theatre.

[61] *Indian Talkie, 1931–56,* list, pp. i–xxvii.
[62] *Indian Cinematograph Year Book, 1938,* p. 14.
[63] Marred, to be sure, by some discordant notes. A Bombay trade paper asked Gandhi for a message of congratulation to the film industry on its anniversary, and received this message from the Mahatma's secretary: "As a rule Gandhi gives messages only on rare occasions—and these only for causes whose virtue is ever undoubtful. As for the Cinema Industry he has the least interest in it and one may not expect a word of appreciation from him." Reported in *Dipali,* June 16, 1939.

Inevitably new interests and needs grew into new departments. Bombay Talkies maintained a school for children of staff members, which also became a school for child actors. Prize possessions of the costume department became a "museum" of historic costumes. Books acquired for reference became a "library" of 3,000 books and manuscripts. Prabhat had its "zoo," including tigers, deer, and birds.[64] Prabhat also had a swimming pool, for recreation as well as production needs. Bombay Talkies had its own physician, who operated a clinic and also supervised the sanitary practices of the canteen, which served breakfast, lunch, dinner, and—for scene builders—midnight snacks.

The educational impact of organizations of this sort was considerable. New Theatres required its staff members to be on hand every day during working hours, whether or not there was an assignment. When not acting, an actor might be put to fencing or riding lessons. Or he might be given temporary technical duties. At Bombay Talkies an actor was expected to do some work as a cutter, as an essential part of his film training. Similarly, a technician might occasionally perform.

Although some performers were "stars" in that they were widely known and featured in publicity, no real star system had as yet developed. The star was an employee; he or she was not the pivot of planning and was not in control.

Producer and director were the dominant figures. Throughout the 1930s the difference between the salaries of top actors and other actors remained small by the standards of later years. Throughout this period Rs. 3,000 per month remained the ceiling for star salaries at several of the larger companies.[65] An established lesser actor might get Rs. 600; a beginner, Rs. 60.

Looked at from the vantage point of later decades, the companies of the 1930s carried a large "overhead" organization. In truth, the self-sufficient studio could exist only because of the low salaries that were considered acceptable. Basic salaries in India were of course kept low by the existence of tens of millions living on the edge of destitution. In addition, the 1930s were a time of depres-

[64] Fathelal, "Prabhat Was a Training School," in *Indian Talkie, 1931–56,* p. 139.
[65] *Indian Talkie, 1931–56,* pp. 127, 143.

sion. The security offered by the one-big-family studios outweighed all considerations of level of pay.

One of the most fervent architects of the one-big-family company was Himansu Rai. He supervised to the last detail the planning and construction of the Bombay Talkies studio and other buildings. Selection and training of personnel received the same minute attention. Celebrated authors and scholars were constantly enlisted to conduct personnel seminars. Assignment of staff workers to a variety of duties that would broaden their conception of the film medium was a policy he personally implemented. He planned recreational facilities, took interest in health problems, and even supervised the purchase and installation of electric stoves for the canteen.[66]

In fact, Himansu Rai was incapable of sparing himself. In 1940 he had a nervous collapse, followed quickly by pneumonia, and was taken to a hospital. While he lay ill a new script was completed.

[66] Interview, Devika Rani Roerich.

A BIRTHDAY PARTY AT BOMBAY TALKIES FOR DEVIKA RANI, CA. 1938

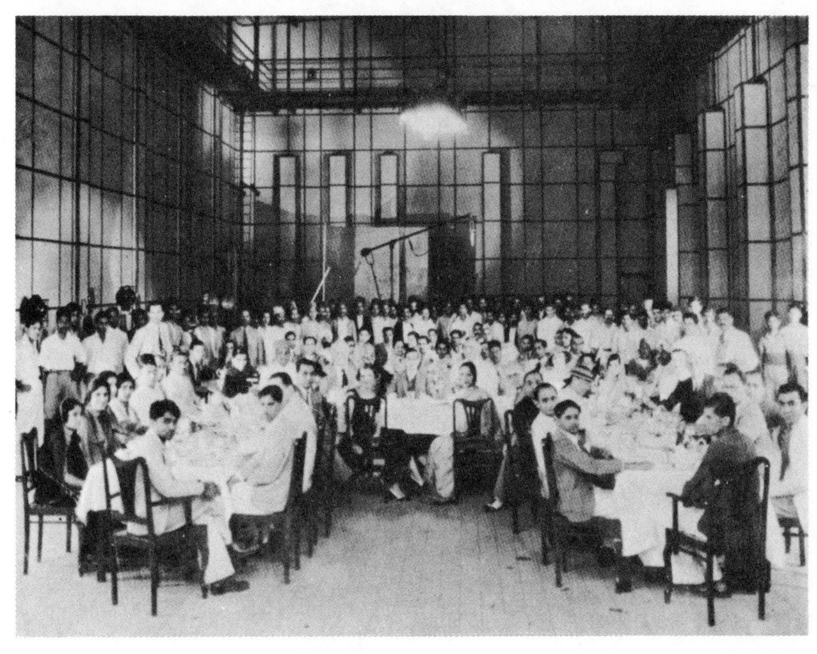

Because the final version, the "blue book" copy, always needed his approval, it was brought to him at the hospital. He could scarcely hold it. Clutching it to him for a moment, he gave it a sort of blessing. He died soon afterward.

At this time the one-big-family studio must still have seemed a secure, well-established institution. Yet the elements that would destroy it were very much at work.

The film business in 1940 was prosperous enough to attract new entrepreneurs, new capital. As in the First World War, defense industries were putting new money into circulation. The result was an influx of new producers. Most of them had no studio—but studios could be rented. They had no laboratory—but such services could now be purchased. They had no acting staffs—but this too could be solved, with sufficient funds.

These new independent producers, seeing the crowded theatres, guessed that the idolized stars were the key to financial triumph, and began to make offers to stars on a per-picture basis. A star suddenly found he could earn more in a one-picture contract than he had been accustomed to earn in a year of employment. Stars began to leave the big companies. The institution of free-lancing—which had existed on a very minor scale—grew rapidly. Directors and song writers were similarly lured by large fees into a free-lance life.

In the early years of the war the big studios began to find their self-sufficiency ebbing. They sometimes had to bargain competitively for stars—and then wait their turn. By the end of 1941 various stars were found to be "making three or four pictures simultaneously."[67] As a result, production schedules were slowed. Under these circumstances the old companies could not afford to maintain large full-time staffs. They began to divest themselves of overhead.

In 1941, of the 61 films produced in Bombay, only 40 came from established producers. Newcomers—*without* studios, *without* laboratories, *without* acting staffs, *without* libraries, *without* canteens, zoos, museums, schools—had produced 21 films.[68] The big companies were losing their hold. As their position weakened, an exodus began.

[67] *Journal of the Film Industry*, February, 1942.
[68] *Ibid.*, January, 1942.

In 1941 Shantaram left Prabhat to "produce under his own banner." In 1942 one of Sagar's leading directors, Mehboob R. Khan—who had directed its *Jagirdar*—began producing "under his own banner." This process was briefly arrested by a war shortage of raw film. A government allocation system, favoring "established producers," gave them a temporary respite. During 1944 and 1945 the one-big-family studios lived on, uneasily, with the aid of this special protection. Then the fragmentation began again. By the end of the decade the one-big-family studio was an extinct species. Those that survived in name had substantially changed their structure.

The production statistics of the 1940s tell the story of the decline and fall of the one-big-family studio. The decline was paralleled by the rise of the independent producer—independent of overhead—who used a rented studio and free-lance talent. The new producer might be a complete outsider, or a star or director who had found a backer. Many a studio owner, renting his studio to these new producers, gradually became dependent on them—and so helped to entrench the new system and destroy the old. Here are the figures:[69]

	New producers releasing films during the year	*Total number of producers releasing films during the year*	*Number of films released during the year*
1940	42	100	171
1941	46	103	170
1942	55	108	163
1943	46	110	159
1944	28	95	126
1945	10	84	99
1946	66	151	200
1947	125	214	283
1948	126	228	289
1950	113	197	241

But these are only the statistics. The shift was not only one in business practices. Behind the figures lay vast changes in the climate of an industry. To understand them fully, we must look beyond them to other changes—in India and the world.

[69] *Indian Motion Picture Almanac and Who's Who, 1953,* p. 238.

A State of War

For India, as for all the world, the 1930s were a time of mounting tension. Throughout the decade there was the sense of coming world struggle. Wars in Manchuria, Spain, and Ethiopia, fiercely fought as they were, seemed only rehearsals for some mightier Armageddon, in which the world powers of the West would be, once more, under mortal strain. In India the conviction grew that a decisive moment of history was in the making, which would yield to India her freedom.

Tragically, this confidence brought with it rising bitterness. As deliverance seemed nearer, the structure of the future independent India became an increasingly burning issue. At the start of the decade, the differing views of the Indian National Congress, the All-India Muslim League, and other groups still seemed to offer hope of conciliation. By the end of the decade this was no longer so. Hindu-Muslim tensions were on the increase. The idea of a separate Muslim state, a "Pakistan," had been discussed since 1933 and by the end of the decade presented itself as a holy crusade. At the same time the charge was increasingly heard that the British, in encouraging Muslim sensitivities and fears, were using the issue to "divide and rule."

In 1935 Great Britain passed an India Bill, adopting reforms to go into effect two years later. Burma would acquire a separate administration, which would place her close to dominion status. In India the reforms would provide for an extended franchise and transfer of various administrative powers from British to Indian hands. But the Viceroy in the central government, and the British-appointed governors in the provinces, would retain wide powers: independent legislative power, veto power, and emergency powers. Under these the executive could, for example, take control in any matter involving minority rights—which in the eyes of the Congress leaders, gave the British a vested interest in minority disaffection. In the Congress the prevailing attitude was that of Jawaharlal

Nehru, who called the reforms "a new charter of slavery."[1] When the Congress decided to take part in the elections under these reforms, its initial purpose was to obstruct the new government.

Meanwhile left-wing influence in the Congress was growing. One of its most-admired leaders, Subhas Chandra Bose, was among those who unceasingly demanded more militant measures toward the winning of independence. As early as 1936 the Congress, under pressure both from radical factions and pacifist, gradualist factions, was warning Great Britain that India would not again participate in an "imperialist war."

Great Britain meanwhile went ahead with its reforms. In 1937, with the inauguration of the new government, and in an effort to promote an atmosphere of harmony, the British allowed release of a number of topical films which had long been banned—some since 1930. The list was an astonishing one, suggesting the extent of British determination throughout the decade to keep the passions of independence out of the film medium. The following films were released:[2]

1. *Mahatma Gandhi's March for Freedom* (Sharda Film Co.)
2. *Mahatma Gandhi's Historic March, March 12, 1930* (Krishna Film Co.)
3. *Mahatma Gandhi's March, March 12, Ahmedabad* (Ranjit)
4. *Mahatma Gandhi's Return from London* (Krishna)
5. *Topical of Mahatma Gandhi and Others* (Indian Topical Co.)
6. *Bombay Welcomes Mahatma Gandhi* (Billimoria)
7. *Bombay Welcomes Mahatma Gandhi*—with vernacular subtitles (Billimoria)
8. *The Return of Mahatma Gandhi from the Round Table Conference*—synchronized (Imperial)
9. *The Return of Mahatma Gandhi from the Round Table Conference* (Imperial)
10. *Mahatma Gandhi's Speech in Public Meeting*—synchronized (Krishnatone)
11. *Mahatma Gandhi Returns from the Pilgrimage of Peace* (Saraswati)
12. *Forty-fifth Indian National Congress at Karachi*—with Gujarati titles (Eastern Film Co.)
13. *Forty-fifth Indian National Congress at Karachi* (Eastern Film Co.)

[1] Quoted in Wallbank, *A Short History,* p. 178.
[2] *Journal of the Motion Picture Society of India,* August, 1937.

14. *Mahatma Gandhi after his Release* (Naujuvàm)
15. *Mahatma Gandhi after the Truce*—synchronized (Imperial)
16. *National Flag Hoisting and Salutation Ceremony* (Bombay Provincial Congress Committee)
17. *Epoch-making Voyage of Mahatma Gandhi to London* (Saraswati)

Here, in capsule form, was a chronicle of the independence drive in earlier years of the decade. The period had begun with an historic gesture in which Gandhi, after a "march to the sea," had entered the surf, dipped up salt water, and, while thousands roared approval, placed it on a fire. The gesture was a symbolic defiance of the governmental salt monopoly and the hated salt tax. Throughout India it had spurred resistance to tax collection and made such resistance a sacred mission. It had led to riots and eventual imprisonment of 60,000 Congress members, including Gandhi.[3] All this and later symbolic acts and drastic repressions were recalled by the release of the long uncertified films.

The easing of censorship proved only momentary. As Congress leaders made increasingly clear their determination to boycott the coming war, British censorship tried to maintain a blackout of the Congress. Film producers now took to the casual introduction of Congress symbols into films. On a wall, in the background, one would see the Gandhian motif, the spinning wheel, signifying defiance of the economic pattern of empire. In a store there would be a calendar with Gandhi's portrait; in a home, a photograph of Nehru; on the sound track, the effect of a passing parade, with a few bars of a favorite Congress song. Often such symbols had no plot reference; but in theatres they elicited cheers. As war began, British censors ordered the scissoring of such shots. After 1942, when Gandhi was again imprisoned—along with a number of other Congress leaders—no photograph of Gandhi was allowed on the screen, no matter how incidentally. *The Journal of the Film Industry* eventually advised the government: "Excision of photos of the Congress leaders is not going to remove them from the hearts of their followers."[4]

Throughout these years of repressive censorship, Great Britain had also made varied efforts to win Indian public opinion. In the

[3] Wallbank, *A Short History*, p. 169.
[4] *Journal of the Film Industry*, February, 1944.

early 1930s it launched an Empire short-wave radio service from London. In 1935 a British Broadcasting Corporation producer, Lionel Fielden, was dispatched to "upgrade" the government-operated Indian radio system. One purpose was to broaden the potential audience for the London short-wave programs beamed to India.

On September 3, 1939, after the Nazi invasion of Poland, Great Britain declared war on Germany; on the same day the Viceroy of India, Lord Linlithgow, declared India to be at war with Germany. The Indian National Congress, condemning the resolution, said that such action could only be taken by the Indian people. The Congress expressed its abhorrence of "Fascism and Nazism and their glorification of war" but added: "India cannot associate herself in a war said to be for democratic freedom when that very freedom is denied her . . . a free democratic India will gladly associate

CONGRESS SYMBOLS IN ACTION: *Seva Sadan* (SERVICE HOME), 1938, SPINNING WHEEL NUMBER. MADRAS UNITED ARTISTES AND CHANDRAPRABHA (TAMIL)

herself with other free nations for mutual defence."[5] The Congress continued resolutely to boycott the war effort.

Most leading film producers supported the Congress position. But some, moved by reports of Nazi brutality and apprehensive over the German-Japanese alliance, were beginning to have other thoughts. When the government formed a Film Advisory Board in 1940 to make, and encourage the production of, war-effort films, J. B. H. Wadia, producer of stunt films and newsreels, readily accepted the chairmanship. British documentary film specialists—Alexander Shaw and later others—were sent from London to reenforce the work. Theatres were at first urged, not compelled, to show these films. But distribution results were very unsatisfactory and in 1943 the showing of war documentaries, either produced or "approved" by the government, was made compulsory. Meanwhile the government had also become involved in newsreel production. In 1940 it had signed with Twentieth Century–Fox a contract under which the latter made Indian-language dubbings of the British *Movietone News*. This later developed into a special newsreel keyed to Indian interests, a government-sponsored *Indian Movietone News*—which the government took over completely in 1943 and renamed *Indian News Parade*.

The showing of *Indian News Parade,* as of the documentaries—meanwhile renamed *Information Films of India*—became compulsory. Theatres also had to pay for the films; the rates ranged from Rs. 2/8 per week to Rs. 30 per week.[6] The private production of topicals now almost completely stopped. Great Britain thus put into effect a pattern of operation that was to play an important and controversial role in independent India.

But wartime developments in censorship, and in the compulsory showing of government films, may have had a less momentous effect on the film industry than other, simultaneous developments. These were perhaps not much noticed at the time, but their effect was far-reaching.

[5] Quoted in Azad, *India Wins Freedom,* pp. 26–30.
[6] *Journal of the Film Industry,* June, 1943.

In black and white

The year 1940 brought to India a spurt in industrial activity. Iron and steel production was expanding. Indian factories, in spite of the war boycott by the National Congress, were making field guns, machine guns, bombs, depth charges, and ammunition for British and Allied forces in various parts of the world. Increased employment put extra money in circulation. The motion picture theatres were crowded. A drift from rural areas to city factories augmented the boom. Meanwhile the industrial growth also brought shortages and, as in the First World War, a black market in essential items such as steel, cement, cotton, foods. Anticipation of rising prices brought speculation. When rice prices shot up, fortunes were made by the rice speculators. Thus the bulging funds in circulation included not only the wages of industrial workers but also various kinds of illicit profit. The new money that became available to the film industry included black market money.

A problem for the black marketeer was that his profits could not be openly reinvested. Therefore offers made to film stars in the early 1940s included a device that was apparently new to the film world. The star would receive a one-film contract calling for payment of Rs. 20,000. In actuality he would receive Rs. 50,000, but the additional Rs. 30,000 would be in cash, without any written record. To the star this extra sum, this payment "in black," was of course tax-free. This not only made it especially attractive but gave it a patriotic tinge. For years the withholding of taxes from the British Empire had been held a service to freedom. Now, when taxes were going into an "imperialist war," denounced and boycotted by the National Congress, evasion of taxes was all the more easily rationalized—if need for rationalization was ever felt. The star's delight, personal and patriotic, in a partly tax-free salary coincided neatly with the investor's interest in off-the-record investment.

Not surprisingly, the "black payment" system soon spread, though on a smaller scale, to other key figures such as music directors[7]—considered, after the stars, the most important element in

[7] The music director, in Indian film parlance, is usually a composer-arranger-conductor. He generally works with a lyricist or "poet."

box-office success. To receive part of one's salary "in black" was a badge of distinction. The rumor that this or that star received 75 percent of his salary "in black" and only 25 percent "in white" came to be heard frequently in trade circles, and contributed to the star's prestige and bargaining power.

As the industry became more and more fragmented into small production units, which more often than not dissolved after one production, to be replaced by others, the film industry became an increasingly attractive investment opportunity for black marketeer and profiteer. Just how large a role this played in the rise of the "mushroom producers" in the early war years and after the war, no industry or government statistics can tell us. We shall find an enquiry committee of independent India concerning itself in later years with this thorny and persistent problem.

It was not only the stars and other key figures whose lives were changed by the inflationary war period. During most of these years, controls over scarce materials curtailed construction of theatres. While active producers multiplied, there was no corresponding increase in exhibition outlets. The exhibitor was now subsidized by scarcity of competition. The days when an exhibitor feared he would not have films to show were gone. Fear had shifted to the producer: would he have an outlet? As the producer was subjected to more and more competition and pressure, power shifted to distributor and exhibitor.

These knew what they wanted, just as the financier knew what he wanted: big star, eight hit songs, several dances. Producers, to clinch investments and distribution, knew that these were the fixed essentials. They therefore found themselves bidding competitively—and suicidally—for the small group of "big" stars so designated by distributors and exhibitors. As star fees shot up from Rs. 20,000 to Rs. 75,000 to Rs. 100,000 to Rs. 200,000 per film, the producer's own position grew more precarious. Yet the alternatives, as many saw them, were to play the game or quit. From a prewar average of Rs. 90,000,[8] production budgets in Bombay jumped to a postwar range of Rs. 400,000–500,000.[9] It was not unusual for stars

[8] *Journal of the Motion Picture Society of India,* January, 1937.
[9] *Journal of the Film Industry,* April 1950.

to receive half the budget. Failure meant catastrophe for the producer.

The producer, under increasing pressure, economized where he could. Training programs became an extinct notion. The lesser actors and extras, set adrift to free-lance, were paid at the old wages, but only when they were needed. Writers also felt the squeeze. Beverley Nichols, surveying the Bombay film world in 1944, found its stars "actually better off" than Hollywood stars, since their taxes were "a fleabite"; but the writer of a feature film "considers himself lucky to receive two hundred rupees." He adds: "That is one reason why Indian films are marking time."[10]

For most people in the industry, earnings and security were skidding. So were values, but during the war years the film world did not yet think of the change in these terms. The full effect would only be seen later. During the war years there was an impression of dynamic activity. There were more companies at work. Money was available. Films were being made in a somewhat different business framework, but many of the same directors, actors, and technicians were at work on them. Many of the films had the same essential elements as in prewar years. Surely many of them were excellent. It was only in later years that people became aware of the changes that had taken place.

In 1947 the *Journal of the Film Industry* looked back on the war years:

> With black markets and corruption abounding in the country, businessmen began to think in terms of easy money and quick returns. . . . The inflationary war boom has been the greatest encouragement for all and sundry to enter the various branches of the film industry in India.[11]

A screen writer who had been with Prabhat and Bombay Talkies looked back at the same years:

> As the industry gained dimensions, mushroom producers came in large numbers, and the first thing they did was to eliminate the position of the writers. They wrote stories themselves or adapted Hollywood films. After the Second World War, making of original stories became taboo.[12]

[10] Nichols, *Verdict on India*, p. 110.
[11] *Journal of the Film Industry*, September, 1947.
[12] Vyas, "Writers Were Better Respected," in *Indian Talkie, 1931–56*, p. 119.

A feminine star of New Theatres, who had played opposite Prince Barua in many films, looked back at the same years:

> The "star system" met with great success in the abnormally strained atmosphere of the war. It grew in size and in the huge deluge that it produced, the producer, the technicians, the writer and others were washed away.[13]

A producer, founder of Sagar Movietone, also looked back:

> After the last great war . . . stars asked fabulous prices and on top of it, did not agree to work exclusively. So no other course was left to us but to close down and so we did.[14]

Allocation

In the later years of the war there was still another development that must be mentioned. Like other problems arising during the war, this also proved to be an augury of problems to come in the years of independence.

In 1942 the mushrooming of production resulted in a shortage of raw film. The government temporarily eased this by putting a limit of 11,000 feet on the length of feature films. By the following year more drastic action was needed and a government allocation of raw film was begun. This placed a new weapon in government hands.

We have seen that government policy favored "established producers" and that this granted a temporary reprieve to the old, large production units. But allocation had other aspects that especially troubled these same producers. To receive regular allocations, the government decided that a producer must devote at least one feature film in three to a "war-effort" theme. After two nonwar films, he would have to submit plans for a war-effort film in order to receive a continuing supply of raw stock. Most producers and directors, as we have noted, supported the Congress noncooperation policy. Studio-owning producers therefore faced a painful dilemma. Professional survival required some activity acceptable to the British as war effort.

The dilemma produced some catastrophic films. Bombay Talkies, where Devika Rani was now in charge of production, secured

[13] Devi, "Rise of the Star System," *ibid.*, p. 135.
[14] Desai, "Sagar Was Ten Years Ahead," *ibid.*, p. 127.

approval of a film on nurses as a war-effort film. Titled *Char Anken* (Four Eyes), it was a total failure at the box office, the first such disaster for Bombay Talkies. Similarly, New Theatres produced *Dushman* (Enemy), on tuberculosis. It had a better reception but did not add to the glory or resources of New Theatres.

The astonishing military advances of Japan throughout 1942 brought some shift in opinion, which helped a few producers to solve the war-effort problem. That year Japan took Singapore and swept through all of Burma. Suddenly Japan stood at India's doorstep. To some Indians the Japanese army represented a force of liberation, embodying the idea of "Asia for the Asians." Subhas Chandra Bose, former Congress president, fought with the Japanese forces, leading an Indian National Army and giving strength and substance to Japanese propaganda. But to some Indians the Japanese seemed a threat rather than a force for freedom. This created for more than one producer a war-effort opportunity and led to at least one comedy of errors.

K. Subrahmanyam, founder of the Madras United Artistes Corporation and a member of the National Congress, submitted proposals for a war-effort film, in the Tamil language, dealing with the imminent might of Japan. In so doing, he was condemned by some Congress members. The film seemed to tell the Japanese—"the enemy"—that India did not need her and could take care of her own affairs. Titled *Manasamrakshanam* (In Defense of Honor), it was approved and released, and became an unexpected box-office success. Seeing the spirited audience reaction, authorities began to have some uneasiness about the film. They became aware that its message could be interpreted in more than one way. In saying, "India can take care of herself," it could be addressing the British as well as the Japanese. In saying, "Go away," it could be echoing the Congress slogan of the hour, "Quit India!" In a few towns, local authorities decided to uncertify the war-effort film. And Subrahmanyam was regarded as a Congress stalwart rather than an outcast.

At the start of the war, those who wished India to follow in the path of the Soviet Union strongly opposed participation in the "imperialist war." They urged, along with Subhas Chandra Bose and other militants, a vigorous course of action for independence.

After Hitler invaded Russia, there was a sudden split among the militants. Some, like Subhas Chandra Bose, continued to urge militant action against the British raj; others were now urging cooperation in the war effort with Britain and Russia and their allies. This realignment of forces helps to explain what was the most interesting and successful of war-effort films, *The Journey of Dr. Kotnis.*

Its author and instigator was a rising young journalist of vigor and skill, Khwaja Ahmad Abbas. In *The Journey of Dr. Kotnis* he demonstrated an astute comprehension of political forces that was to make him an increasingly prominent figure, one who often played an international role in film.

Throughout his career he was a journalist with a foot in the film field. After graduating from Aligarh University, he got a job on the Bombay *Chronicle,* and also did part-time publicity work at Bombay Talkies. At the *Chronicle* the aging film critic sometimes let Abbas review films for him. When the critic died in 1938, Abbas inherited the position. He soon became one of the most unpopular figures in the Indian film world. His pungent, lively criticisms brought to the *Chronicle* threats of an advertising boycott from film interests. The *Chronicle* is said to have solved this dilemma by elevating Abbas to the editorship of its Sunday edition.[15] The sequence of events stirred Abbas to write a film script, *Naya Sansar* (The New World), about a journalist under pressure from business tycoons—which Bombay Talkies undertook to produce. Made in Hindi, released in 1941, the film was a popular success and launched a vogue in "new world" films with "progressive" themes. It was against the background of these rising fortunes that Abbas approached V. Shantaram, now working in Bombay as an independent producer, with a suggestion for solving the war-effort problem.

The National Congress, while maintaining a policy of noninvolvement in imperialist war, had expressed its sympathy for the Chinese by sending a medical mission to China. Nehru had sponsored this humanitarian project, in which seven Indian doctors had served with the Chinese fighting forces. One doctor had married a Chinese nurse and eventually lost his life while on duty. Abbas had

[15] Interview, Abbas.

WAR-EFFORT FILM: *The Journey of Dr. Kotnis,* 1946. RAJKAMAL KALAMANDIR (HINDI, ENGLISH)

written a short book about him, *And One Did Not Come Back,* and now suggested that Shantaram produce a film on the same topic. Shantaram accepted the suggestion, and Abbas participated in the writing of the script.

Because the story was anti-Japanese, the British promptly approved *The Journey of Dr. Kotnis* as a war-effort project. Because it dealt with a Nehru-sponsored mission of mercy, the Congress promptly applauded. Because the Chinese force depicted in the film was the famous communist Eighth Route Army of Mao Tse-tung—which eventually established communist China—the Communist Party likewise applauded. The film even aroused American interest: after the war a contract was signed with Mayer-Burstyn, a distributing company specializing in foreign films, for United States art-theatre distribution of *The Journey of Dr. Kotnis.* Abbas had hit on a theme that could cement divergent interests. We shall find him in later years demonstrating the same acumen in the organizing of a USSR-Indian co-production, a diplomatic coup of the 1950s.

The Journey of Dr. Kotnis had one more facet of interest. In anticipation of possible markets in the English-speaking world, particularly in the United States, V. Shantaram decided to make the film as a Hindi-English double version. The Hindi title was *Dr. Kotnis Ki Amar Kahani.* Shantaram himself played Dr. Kotnis in both versions, doing the successive takes in Hindi and in fairly acceptable English.[16] However, with a curious added twist illustrative of Indian preconceptions about the American market, Shantaram decided there should be a change of clothes before the English-language takes. To please American audiences, he felt the clothing should be "more Indian" in the English-language version. The film was therefore given a tourist picture-postcard quality. Perhaps American films on India had persuaded Shantaram to adopt this strategy.

On the whole, however, with the exception of *The Journey of Dr. Kotnis,* British efforts to use raw-film allocation as a lever to elicit war-effort films were not a conspicuous success.

[16] He has been criticized for speaking Hindi with a Marathi accent.

New Day

It is not surprising that the film industry, like the rest of India, looked on the coming of independence as the solution to most of its problems. The disposition to ascribe all problems to the British was widespread.

In Great Britain, shortly after V-E Day in 1945, the Labour Party came to power. Clement Attlee, the new prime minister, announced that his administration was determined to have Great Britain leave India by 1948. The departure was, in fact, accomplished a year earlier—by mid-August, 1947. In the intervening period India was in turmoil. Because of continued inflation and rising costs, the real wages of most people had declined, and millions suffered hardship. In 1945 there was a drought in Bengal, and thousands died in famine. In 1946 there were strikes and riots in many parts of India, and violent attacks on foreign-owned businesses. And as the partition of India became inevitable, Hindu-Muslim bitterness also increased.

On July 4, 1947, an Indian Independence Bill was offered in the British Parliament and speedily passed. On midnight of August 14 India and Pakistan became independent nations. Amid wild jubilation, the *Journal of the Film Industry* in its August issue declared:

> . . . two centuries of economic and political serfdom have come to an end. . . . Under foreign yoke India's ancient art and culture were ignored. Base imitation of alien customs, manners and culture took the place of any organized effort to develop the indigenous art and culture. If the exit of British rule means anything at all this has to change and change thoroughly.[1]

The rejoicing was cut short almost at once by shocking events. The September issue of the *Journal of the Film Industry* opened with these words: "The first month of our freedom has set back the clock of our progress."[2] As refugees by the millions—Hindus and Sikhs leaving Pakistan for India, Muslims leaving India for Pakistan—streamed through the border areas in opposite directions, there had been murderous clashes. As terror spread, the massacres grew more

[1] *Journal of the Film Industry,* August, 1947.
[2] *Ibid.,* September, 1947.

AUGUST 15, 1947:
EUPHORIA

FEBRUARY 12, 1948:
DISMAY

ADVERTISEMENTS, *Times of India*

grisly. Nehru, now leader of a nation, cried out: "Is this the realization of our dreams of a free India?"[3]

Ceaselessly, Mohandas Gandhi pleaded and fasted for an end of violence. He pleaded, too, that Indian leaders pledge themselves to oppose any anti-Muslim program, and pledge also to protect with all their power Muslim lives and property. Nehru and others took such a pledge. Slowly, very slowly, the fury subsided, but not until it had reached a tragic climax. On January 30, 1948, Gandhi was shot to death by a Hindu fanatic, Nathuram Godse. The body of the frail, revered leader was cremated near Delhi; part of the ashes

[3] Quoted *ibid.*

was buried there, and other parts dropped symbolically into the waters of the Ganges and on the Himalayan slopes.

Sobered by the sequence of horror, new India began to tackle its problems as a nation. Besides the settling of 1,250,000 homeless refugees, there were the problems of the integration of several hundred princely states, disputes over Kashmir, a food crisis, the drafting of a new constitution, and the planning of long-range development and social reform. The new nation was 18 percent literate. Average life expectancy among its people was twenty-six years. The range of problems to be attacked—in agriculture, education, health, industrial growth—was staggering.

It is not surprising that both central and provincial governments began at once to look for new sources of revenue, and that they soon took note of the film industry. This industry seemed, at least to the casual observer, to be in a glittering state of prosperity. Its stars were rumored to have astonishing salaries, and a few had begun to live like a new species of maharaja. The old-style maharajas were now being "integrated." Most were quietly relinquishing their power and being pensioned. Even the Nizam of Hyderabad, after armed conflict, became as other men. But a new symbol of glamor and affluence was rapidly rising to take their place: the screen star.

Many of the leaders of the new India were studious, ascetic men. Many were products of a rigorous education. Some had spent long years in jail, reading ceaselessly in the history of the world, pondering the rise and fall of nations, and planning the coming transformation of India. They were intent on a vast program of change. As for the film industry, they had, like Gandhi, "the least interest in it." The idea of entertainment as a necessity of life was not familiar to them. If they thought of film, they thought of it as a potential instrument of social reform that was not being used in that way. They thought of it as too much involved with romance and immature hero worship. They associated it with Western influences that needed to be purged. They also saw it as a source of revenue. Not surprisingly, the year 1947 inaugurated a long series of measures affecting film that soon left the film world dazed.

Before the war most provinces had had entertainment taxes of 12½ percent. There had been wartime increases, considered temporary. But on the heels of freedom came a wave of further increases. By 1949, when the country was being divided into "states," the entertainment taxes ranged from 25 percent to 75 percent, with an average of 33½ percent.[4] It must have been a disturbing experience for Indian exhibitors to read in June, 1949: "BRITAIN EXEMPTS 677 CINEMAS FROM ENTERTAINMENT TAX."[5] The headline referred to rural cinemas in the United Kingdom, which the British government was trying to foster through special assistance.

In India, increased state taxes were only the beginning. Some municipalities also began to levy entertainment taxes.[6] Others developed an ingenious new levy

> for placing policemen on cinema fronts . . . as if the burden of maintaining peace before the cinemas rested on the owners of the cinemas and not on the government, as if the cinema audience did not compose of tax payers who had the right of protection wherever they were.[7]

These levies were in addition to taxes already in existence. Most municipalities were levying octroi duties on the transport of films from one place to another. There were also sales taxes, under which basic cinema equipment was taxed at luxury rates. Some of the taxes, such as internal customs duties, seemed like obsolete remnants of a previous era.[8] And of course there were income taxes—business and personal—and import duties on raw film and production equipment.

There were still other charges, not called taxes, but which seemed like taxation to the film industry. The British requirement that theatres must show government-approved documentaries and newsreels had become inoperative in 1946. It had lapsed because National Congress representatives in the central legislature had succeeded in reducing the government film appropriation to one rupee. The nationalist leaders, on the verge of winning independence, regarded Information Films of India as "a dreadful institution" which had helped to dragoon the nation into war.[9] They thus took the first opportunity to annihilate it. But scarcely a year later, in October,

[4] *BMPA Journal,* July, 1949. [5] *Ibid.,* June, 1949.
[6] *Report of the Film Enquiry Committee,* p. 36. [7] *BMPA Journal,* June, 1949.
[8] *Report of the Film Enquiry Committee,* p. 205.
[9] Mohan, "Panorama of the Private Sector of Indian Short Film Industry," *Marg,* June, 1960.

1947, they decided to revive the mechanism in almost identical form. The states were asked by the central government to insert in all theatre licenses the requirement that a minimum amount of "approved" film be included in every program. During the following months all states complied with this request. Meanwhile a governmental Films Division, on the model of Information Films of India, was launched to produce such films. It promptly presented to all theatres a 52-week block-booking contract under which the Films Division would supply the obligatory films, and the theatre would pay for them. But whereas the British had charged Rs. 2/8 to Rs. 30 per week for providing this "service" to the theatre,[10] the Films Division charge would range from Rs. 5 to Rs. 150 per week.[11]

There were further increased levies. In 1950 the new government decided to institute a Central Board of Film Censors. There would still be censor panels in Bombay, Madras, and Calcutta, but under centralized authority. The film industry, which had long protested "vagaries of censorship," welcomed the move. But shortly before launching the new system in January, 1951, the government announced a new schedule of charges for the censorship "service." Under the British raj the producer had been charged Rs. 5 per thousand feet to have a film reviewed by the censors; the rate would now be Rs. 40 per thousand feet—an increase of 700 percent.[12]

There were still further new costs. A Calcutta producer sending a Bengali film to East Pakistan, his only market outside the new India, now found that in order to get his print back he would have to pay an import duty to his own government, far exceeding the value of the print.[13] The exporter of a Hindi-language film to West Pakistan had the same problem.

By midsummer, 1949, the film associations were estimating that 60 percent of all box-office receipts were going into taxes of one sort or another.[14]

If film makers thought that censorship would be relaxed with independence, they were mistaken. The new regime was moved, at

[10] *Journal of the Film Industry,* June, 1943.
[11] *Report of the Film Enquiry Committee,* p. 24.
[12] *BMPA Journal,* January, 1951.
[13] *Report of the Film Enquiry Committee,* p. 121.
[14] *Journal of the Film Industry,* July, 1949.

every level, by a reformist zeal. Even before the actual date of independence, with Indian-controlled governments already at work in the provinces, reform moves began. The provincial government of Bombay, having decided to introduce prohibition, announced: ". . . with effect from the 1st of April, 1947, no scenes showing drinking of any type of liquor will be permitted in any films."[15] A barrage of protests, letters, and delegations brought a clarifying order: "There are films which are avowedly meant to propagate the idea of abstinence. In such films, drinking scenes being meant to condemn drink will not be cut out."[16] But the film world was aghast that the original ruling could have been issued without any consultation or hearings.

The reform spirit showed itself in other ways. Another Bombay government body, in a report on music education, took occasion to condemn current film songs as seriously corrupted by Western influences and "alien to the genius of Indian music."[17] The reform spirit was further shown in censorship actions, which with increasing frequency cited no existing published criteria. With impartiality, censors began to reject both Indian and foreign films on grounds improvised as needed:

Matlabi—Hindi. Jagitri Pictures. Rejected. This is a sloppy stunt picture, not suitable for public exhibition.

Jassy—English. Eagle-Lion. Prohibited on the ground that the film having no moral behind it is not fit for public exhibition. Trailer is also banned.

The Madonna's Secret—English. Republic Pictures. Prohibited as this is a crime picture without any relieving feature. Trailer is also banned.[18]

As the new centralized board went into action, the trend continued. Increasingly the board seemed to be making aesthetic judgments, as in the following directives, all dealing with the film *Vairabmantra* (Vairabmantra). In each of these rulings, a sequence was ordered to be shortened, without any specific portion being banned. In each of these instances, the censors apparently knew exactly how large a cut would make the sequence acceptable.

[15] *Ibid.*, April, 1947. [16] *Ibid.*, June, 1947. [17] *Ibid.*, July, 1949.
[18] *BMPA Journal,* April, 1949.

Reel 7. *Shorten* the strangulation scene of Sanatan—44 feet.
Reel 8. *Shorten* the journey of the Bhairab to fetch the heroine—193 feet.
Reel 9. *Shorten* the chasing scene—169 feet.
Reel 10. *Shorten* the chasing scene—169 feet.
Reel 12. *Reduce* to the minimum the scene of Tantrik gloating over his plans about the heroine—21 feet. *Shorten* the fire scenes—47 feet.[19]

At the beginning of the independence period, as the new taxes and rulings began, the industry associations took an aggrieved, patient tone. They wrote editorials to explain their grievances, dispatched letters and delegations to various authorities. These seldom produced any results. That "our own government" should have so little interest in the views of the IMPPA (Indian Motion Picture Producers Association), BMPA (Bengal Motion Picture Association), SIFCC (South Indian Film Chamber of Commerce), and other units seemed to them at first difficult to believe. Before long the tone became less patient.

On June 30, 1949, the three above-named associations joined in an All-India Cinema Protest Day, during which virtually all cinemas remained closed as a protest against taxation policies.[20] The associations calculated with some satisfaction how much in taxes had been lost by the central, state, and local governments through the single-day protest. They said: "We may congratulate ourselves on our unity at least."[21] They felt they had perhaps taught the governments a needed lesson, but there is no evidence that the protest day had a restraining effect on any government.

A few weeks later the *BMPA Journal* published an Independence Day anniversary statement:

INDEPENDENCE DAY

Two years ago today, rose the sun over free India's horizon after a lapse of two centuries. That glorious bright morning seemed ever so brighter to India's multitude. They felt happier over the thought that they were free people. High hopes of a better future in the hands of the National Government made them feel happier still. We for ourselves, had felt so. And today, two years after, we find all our hopes shattered. Our own Government takes no notice of our grievances and all our problems remain unsolved. We feel we are on a raft floating on a sea of uncertainties with dwindling rations.[22]

[19] *Ibid.*, April, 1951. [20] *Ibid.*, July, 1949. *Journal of the Film Industry,* July, 1949.
[21] *BMPA Journal,* July, 1949. [22] *Ibid.*, August, 1949.

If the film industry, or its organized leadership, was becoming disenchanted with some of the agencies of free India, it was clear that the latter were not growing in admiration of the film industry, which seemed by now to consist of hundreds of ephemeral enterprises that felt accountable to no one and were engaged in wild and irrational gambles contributing little to the great task of nation-building. If there were, among these implausibly numerous units, individuals who had done fine work and were still trying to do it, they were more and more overwhelmed by the growing chaos. Most of the entrepreneurs were intent on peddling diversions that seemed, at least to the government leaders, trivial and out of tune with the needs of the hour.

Meanwhile the government was itself embarking earnestly on the production of documentaries and newsreels. Asked in the Lok Sabha, the House of the People, whether the government intended to monopolize the newsreel field, or to enlist producers from the private sector, the Minister of Information and Broadcasting, R. R. Diwakar, parried the question: "So far we have been producing these films ourselves. If there are any proposals as mentioned by the member, they will be considered on merits."[23] Meanwhile the government's Standing Finance Committee stated in a report that "it is useless to leave the production of documentaries to private producers."[24] The industry protested furiously.

Thus the film industry and the government began a complex feud that was to last more than a decade. In the background were issues of public versus private enterprise. In the foreground were issues of clashing interests, economic survival, and personality. We shall pursue in later chapters several phases of this long, bitter feud.

But for a time there was a respite, a bright patch in the clouds. Late in 1949 the government announced a new Film Enquiry Committee. At first the industry took a cautious view of this move. But an enquiry committee of exceptional caliber was appointed. It included two representatives of the film industry, both among its notable pioneers: B. N. Sircar, founder of New Theatres Ltd., and V. Shantaram, co-founder of Prabhat Film Company. The committee

[23] *Journal of the Film Industry*, April, 1950.
[24] *Ibid.*, May, 1950.

went to work with a vigor that gradually induced, in the film industry, a restrained expectancy. In its issue of December, 1949, the *BMPA Journal* interrupted its long series of pained protests with an optimistic editorial. The subject was the Film Enquiry Committee and its "earnestness and expedition." The editorial was entitled "A Ray of Hope."[25]

A new enquiry

The chairman of the committee was S. K. Patil, who had been a member of India's Constituent Assembly. The committee held hearings in Bombay, Calcutta, Madras, New Delhi, and six other cities, and studied the 463 replies it received to the 7,140 questionnaires it had sent out. Like the Indian Cinematograph Committee of 1927–28, the committee did its work with dispatch.

Unlike its predecessor, the committee did not publish the testimony it had taken. One reason appears to have been that many witnesses were willing to testify only *in camera*. This seems to have had a connection, at least in a number of cases, with the problem of "black" payments. The committee reported: "Judging . . . from the frequent references to such items of "black" receipts and payments, it would appear that the evil is more widespread than is generally realised and deserves thorough investigation."[26] Testimony on black payments dealt not only with the custom of making such payments to artists but also with off-the-record payments made to theatres in strategic areas, to secure exhibition.[27]

The committee studied the problems of exhibitor, distributor, and producer. Not surprisingly, a major emphasis was on the progressive fragmentation of production and the effect on production quality. The most important recommendations dealt with this problem. The committee dramatized the fragmentation in a series of tables. The years 1946–49 were summarized as follows:

In 1946:

	Total
1 producer produced 7 films	7
1 producer produced 5 films	5
1 producer produced 4 films	4

[25] *BMPA Journal,* December, 1949.
[26] *Report of the Film Enquiry Committee,* p. 93.
[27] *Ibid.,* p. 134.

8	producers produced 3 films each	24
20	producers produced 2 films each	40
120	producers produced 1 film each	120
151	producers produced in all	200 films

Of the 151 producers, 94 dropped out in the following year.

In 1947:

		Total
2	producers produced 7 films each	14
3	producers produced 5 films each	15
2	producers produced 4 films each	8
9	producers produced 3 films each	27
21	producers produced 2 films each	42
177	producers produced 1 film each	177
214	producers produced in all	283 films

The 214 producers included 156 newcomers. Of the 214, 160 dropped out in the following year. Only 58 producers appeared in both the 1946 and 1947 lists.

In 1948:

		Total
1	producer produced 6 films	6
6	producers produced 4 films each	24
4	producers produced 3 films each	12
22	producers produced 2 films each	44
178	producers produced 1 film each	178
211	producers produced in all	264 films

The 211 producers included 157 newcomers. Of the 211, 151 dropped out in the following year. Only 54 producers appeared in both the 1947 and 1948 lists. Only 25 producers appeared in all three lists—1946, 1947, 1948.

In 1949:

		Total
1	producer produced 5 films	5
4	producers produced 4 films each	16
8	producers produced 3 films each	24
29	producers produced 2 films each	58
186	producers produced 1 film each	186
228	producers produced in all	289 films

The 228 producers included 168 newcomers. Only 60 producers appeared in both the 1948 and 1949 lists. Only 18 producers appeared in all four lists—1946, 1947, 1948, 1949.[28]

[28] *Ibid.*, appendix, pp. 323–24.

The newcomers appear to have consisted in part of complete outsiders, investing in film their profits from other fields, and in part of previous producers appearing as new corporate entities. In any event, most of the films in each year were now offered by new units.

The committee recognized the radical changes involved in this:

Within three years of the end of the war, the leadership of the industry had changed from established producers to a variety of successors. Leading "stars," exacting financiers and calculating distributors and exhibitors forged ahead. . . . Ambitions soared high.[29]

The percentage of failure was also high.

Yet such is the glamour of quick and substantial returns which a comparatively small number of producers can secure as a result of the success of their productions that the industry has shown no signs of suffering from lack of new entrepreneurs who are prepared to gamble for high stakes, often at the cost of both the taste of the public and the prosperity of the industry. In the process many of them lose their own private fortunes in a substantial measure, make the general public pay to see pictures which not only discredit their intelligence but also enhance their reputation for credulity and submission to make-believe, and leave the industry "unwept, unhonoured and unsung."[30]

The committee noted that in most cases the producer, whether "established" or new, launched a production without sufficient funds to complete it. This had become customary in the years of fragmentation and rising costs. It meant that at some time during production the producer had to secure new finance, for which he turned either to a distributor or a moneylender. In most cases, one or more distributors became involved during production.

Distributors of Indian films had hardly existed in the 1920s, when many producers dealt directly with exhibitors. But during the sound era distribution had grown in importance and developed a pattern. It had, like production, become fragmented—in 1948 there were 887 distributors in India.[31] The nation was now divided into five distribution territories or circuits. A Hindi film might have some distribution in each of these. External markets regularly reached by Indian-language films, such as East Africa and South Africa, were sometimes treated as a sixth territory. The producer of a Hindi film often dealt with a different distributor for each territory.

[29] *Ibid.*, p. 14. [30] *Ibid.*, p. 16. [31] *Ibid.*, p. 115.

During production the producer generally sold off one or more territories to raise funds to complete the picture. A territory would generally be sold for a lump sum. In the producer's home area, where theatre attendances could be checked, the producer would usually try to get distribution on a percentage basis. If a producer had faith in his film, he would make every effort to retain this territory until after completion of the film. In many cases he would be forced to sell even this during production.

To retain some stake in his production, he might turn to a moneylender for a short-term loan, to tide him over the final weeks or months of production. Loans were generally at illegal rates.

> At present loans are being obtained at rates as high as 60 to 100 per cent per annum. Interest is not paid directly at this rate, but is usually confined to the legitimate figure of 6 per cent or 9 per cent. The lenders, however, charge a "royalty" of not less than 10 per cent on the amount lent, and very often the loan is for a short period of three or six months, the royalty having to be paid again each time the loan is renewed. Royalty and interest are deducted in advance, on each occasion of renewal, making the actual rate of interest very high. The total amount paid within a year for the use of capital thus adds up to an usurious figure.[32]

These suicidal loans were often made necessary by production delays. A major reason for such delays, the committee found, was the involvement of stars in several productions simultaneously.[33]

Producers owning studios and maintaining technical staffs were especially endangered by such delays, but could turn to banks for loans, using their studios as collateral. However, production by studio owners was declining; a number of studios had become "rental studios." The overwhelming majority of producers had no tangible property, could not obtain bank loans, and depended on distributors or moneylenders. This had its effect on production:

> The ultimate necessity of having to sell the picture more or less under duress affects also the quality of films. The producer tends to concentrate on the particular aspects of the picture which would appeal to the distributors and help in securing a quick sale or a good price.[34]

[32] *Ibid.*, p. 96.

[33] The *Report of the Film Enquiry Committee* does not cite details. At the time of the enquiry, some stars appear to have been involved in as many as twelve productions simultaneously; in the mid-1950s, a few are said to have been involved in twenty productions simultaneously.

[34] *Report of the Film Enquiry Committee*, p. 97.

Similarly it has been brought to our notice that distributors make "suggestions" in regard to the story and sometimes about the songs and tunes. Considering the financial relations between producer and distributor, such "suggestions" are generally taken as mandatory by the producer. . . . Distributors appear to have been ultimately responsible for the temporary success of some "stars" who managed to secure on the strength of one "hit" a number of engagements which their merit failed to justify. They appear to have been at least partly responsible for the establishment of certain "cycles" in film-making, resulting in the production of a dozen different variations of a theme.[35]

The committee recognized that it was natural that a distributor should want to safeguard his investment and therefore suggest "what in his view adds to the potential earning capacity of the film."[36] Yet the committee seemed to lean to the view, expressed by some witnesses, that the dominance of the distributor was "baneful," and this led the committee to make one of its most important recommendations.

It was a recommendation made years before by the Indian Cinematograph Committee, and long recommended by the industry. The committee felt that a government-sponsored Film Finance Corporation should be established, with a starting capital of Rs. 10,000,000 and the right to borrow an equivalent sum,[37] to which producers could turn as an alternative source of capital. This corporation, like the similar corporation established in the United Kingdom, would presumably base its decisions on a set of values different from those applied by commercial distributors. The aim was to liberate the producer from the dominance of distributor and moneylender.

The committee felt that such a plan would have to be accompanied by efforts to reduce the chaos in production—to "rationalize" production. It therefore proposed a Film Council of India to "superintend and regulate" the industry.[38] The majority of the council would be representatives of the industry but it would also include representatives of the central government, state governments,

[35] *Ibid.*, pp. 116–17.
[36] *Ibid.*, p. 117.
[37] *Ibid.*, p. 103.
[38] *Ibid.*, p. 187.

and education. As envisaged by the committee, the council would instigate research for the benefit and long-range development of the industry. It would have a bureau for the prior scrutiny of scripts, to curtail unqualified "adventurers." It would have under its supervision an Institute of Film Art for the training of new talent. It would give annual awards for outstanding work in film. It would foster film libraries and film societies, and perform various other functions. The Film Council of India, according to committee recommendations, would be supported by a share of state entertainment taxes, as well as by other taxes including a cess on foreign films; fees for services performed by the council; and nominal contributions by film associations to be sponsored by the council.

In the course of its study, the committee had a look at taxes. While defending a number of existing levies, it agreed with various contentions of the industry concerning the burden of taxation. It recommended that entertainment taxes be made uniform at a level of 20 percent. It criticized such charges as octroi duties, police charges, and internal customs duties.

It supported the obligatory showing of documentary films and newsreels, and the required payments for such films.

It recommended formation of an Export Corporation to stimulate the export market for Indian films. Among other recommendations, the committee proposed that the government embark on a carefully phased program of manufacture, so that India would eventually produce its own raw film as well as various kinds of equipment.

On the whole, the committee had shown a desire to bolster rather than chastise or control the industry. It made clear that many witnesses had proposed nationalization of the film field, or portions of it, but the committee itself opposed such plans. It wished to keep film largely in the private sector. Paying tribute to past achievements, the committee added:

> It is a pity that an industry which has grown to such proportions on its own, without either state support or patronage and in the face of foreign competition on terms which were certainly not much to its advantage, should find itself in the present state of doldrums. It cannot be denied that

the pioneers of this industry established themselves in spite of the adverse circumstances of patronage of foreign goods, of social stigma that attached to the profession, of lack of high quality equipment, and of dearth of artistic and technical talent. Nor can it be gainsaid that the contribution that these pioneers and their successors—famous names in its annals like Prabhat, Bombay Talkies, and New Theatres—made to the building up and growth of this industry despite these adverse circumstances was substantial and praiseworthy. Unfortunately, however, the industry was overtaken by war conditions while it was still none too firm on its feet organisationally, and when the storm and the flood came, it lacked the sturdiness of the giant oak or the strength of the embedded rock.[39]

While emphasizing the impact of war economics on the industry, the committee criticized those in the industry who tended to "ascribe all its difficulties to external factors as opposed to those within the competence of the industry itself to regulate and control."[40]

The committee also noted that witnesses representing the public were freely critical of films but "unrepentant of their cinema-going habits."[41] On the surface, it found the industry in a state of prosperity, with an annual attendance of 600 million people at its 3,250 theatres—including 850 traveling cinemas.[42]

In general, the film associations reacted with delight to the committee report. They had reservations about the extensive powers to be vested in the proposed Film Council of India. But the proposal for a Film Finance Corporation received, as was to be expected, enthusiastic support. The committee's comments on taxation were warmly applauded, and government action was hopefully awaited.

The central government met the report with a long silence. It had other and pressing matters to attend to. A new constitution had gone into operation. In 1951 the world's largest electorate, 107 million voters, went to the polls in its first national election. That same year the first five-year plan, an historic effort to lift India into the twentieth century, was launched. The following year an immense village development program was begun. As to the *Report of the Film Enquiry Committee,* the government said it was studying the matter. Meanwhile the states took no action on entertain-

[39] *Ibid.,* p. 173. [40] *Ibid.,* p. 6.
[41] *Ibid.,* p. 7. [42] *Ibid.,* pp. 14–16.

ment taxes. A year after the submission of the report, virtually no action had been taken on any of its recommendations.

During 1952 the industry became increasingly restive. Overproduction had brought an increasing number of failures. Bengali producers, with 40 percent of their market now in East Pakistan, were having increasing difficulties. The Korean war had brought new inflationary pressures and controls; there was no abatement of black-market problems.

At the end of 1952 the *BMPA Journal* resumed its apocalyptic tone:

> Farewell 1952. The film industry in India will look back on you with regret and remorse. You brought the condition of the industry from bad to worse. The tightening grip of the country's poverty hit us relentlessly. The hopes raised by the report of the Film Enquiry Committee were shattered in the course of your tenure. The Central Board of Film Censors as an expensive Government machinery fed out of the industry's resources forcibly taxed, is a clog to the industry rather than an aid. The Governments —Central and States—instead of helping the industry are trying to help themselves out of the dwindling resources of our industry. . . . may 1953 look on us kindly.[43]

Two months later the *BMPA Journal* ran a short editorial about the report:

> FILM ENQUIRY COMMITEE REPORT
> What has become of that costly document?[44]

The feud was resumed in earnest. A Film Federation of India had been formed in 1951 to represent the industry in its relations with the central government. It now put the tax issue at the top of its agenda.

We shall later trace several phases of this feud—especially struggles over censorship, "approved films," and film music. First we shall have a closer look at each of the main film centers during the first postwar decade, including the first years of independence. Each had by now acquired distinctive characteristics and problems. We shall take the centers in order of their production volume during the decade: Bombay, Madras, Calcutta.

[43] *BMPA Journal,* November-December, 1952.
[44] *Ibid.,* February, 1953.

Industry

As India entered the era of independence, it stood second among the nations in feature production. UNESCO, comparing one-year production statistics, reported the following figures: United States, 459 features; India, 289; Japan, 156.[1]

In India, Bombay was the leading center. Its lead was so commanding that Calcutta was no longer considered, at least in Bombay, a serious rival. Madras still seemed a fumbling beginner. The following table shows the production shares of the three regions in the early postwar years. The figures relate to feature films:[2]

	Bombay area	*Calcutta area*	*Madras area*
1946	77%	12%	11%
1947	70	17	13
1948	65	20	15
1949	60	29	11
1950	60	21	19
1951	54	20	26

It was Bombay that had been affected most by the war boom and the influx of foreigners it had brought. It was especially in Bombay that adventurer producers had injected a new gambling air into the film field. Bombay had become the Indian glamor capital. It was in Bombay that star salaries rose most sharply. Bombay was now almost exclusively concerned with production in Hindi, from which the largest returns were possible.

Among the hundreds of producers active in Bombay in the postwar years, there were some who were continuing from another era. V. Shantaram, who had left Prabhat in 1941 to produce independently, released two features in 1946, two in 1947, one in 1948, two in 1949. He was thus in the select group of only eighteen producers

[1] *World Communications* (1951), p. 26.

[2] *Screen Year Book, 1956,* p. 474. In the table, figures for Poona and Kolhapur are included with Bombay; for Coimbatore and Salem, with Madras. During this period activity in the smaller centers was declining and the dominance of the main centers increasing.

represented in the release lists in each of those years.[3] Each new film from Rajkamal Kalamandir, as he called his independent company, commanded wide attention.

Shantaram had his own way of keeping star problems under control. He played the leading male role in a number of his films, often co-starring with his wife Jayshree. But Shantaram the actor—and his co-star—always received second billing to Shantaram the director. Thus Rajkamal Kalamandir, almost alone among production units, continued to give first place to direction.

In 1946 Shantaram and Jayshree journeyed to the United States, returning the following year. The visit resulted in a contract with an independent distributing organization run by Arthur Mayer and Joseph Burstyn in New York City. With tenacity and dedication, this company was slowly building a United States following for foreign films. Through small "art theatres" in a few cities, they had had success with Italian neorealist films. They now undertook American distribution of selected films of Rajkamal Kalamandir including *The Journey of Dr. Kotnis,* already mentioned, and *Shakuntala.* No Indian film had ever had commercial showings in the United States.

Late in 1947 the following telegram arrived in Bombay:[4]

DEC 5 1947

V SHANTARAM RAJKAMAL STUDIO

DELIGHTED INFORM YOU SHAKUNTALA OPENS CHRISTMAS DAY DECEMBER 25 ART THEATRE STOP ALTHOUGH NOT BROADWAY HOUSE THIS ONE OF FINEST THEATRES NEW YORK SEATING SIX HUNDRED STOP OPENING BEST WEEK IN YEAR STOP TERMS USUAL PERCENTAGE LIKE FOR ALL OUTSTANDING NO GUARANTEE OUR SIDE REGARDS BURSTYN

MARGOLIES

Shakuntala, based on the most famous of the three extant plays of Kalidasa (*ca.* 400 A.D.), had been produced by Shantaram in 1943, in Hindi. Its story, which Kalidása had taken from the *Mahabharata,* tells how King Dushyanta, hunting in the forest, comes on the maiden Shakuntala and rescues her from a bee about to sting her. They fall in love. Dushyanta lingers in the forest and they take their

[3] *Indian Talkie, 1931–56,* list, pp. viii–xi.
[4] Interview and files, Shantaram.

marriage vows. The king must return to court, but Shakuntala is to follow. In simple-hearted devotion she later makes her departure for the distant palace. When she arrives, the king, because of a mysterious curse, has no recollection of her or the marriage. She has meanwhile lost by the river bank the ring the king gave her. The rejected Shakuntala and their child go to live among the beasts of the forest. But the ring, found in a fish, is returned to the king and restores his memory. He goes to find Shakuntala.

The film *Shakuntala,* commanding interest because of its ancient lineage, was praised by some New York film critics, impatiently dismissed by others. It ran twelve days during blizzard weather, then closed. It had no further American bookings. Mayer and Burstyn did not succeed in booking *The Journey of Dr. Kotnis.*

It is possible that *Shakuntala* could not have succeeded in the United States under the best of circumstances. Its story, so rich in as-

SHANTARAM'S *Shakuntala* (SHAKUNTALA), 1943. SHOWN IN NEW YORK, 1947. RAJKAMAL KALAMANDIR

sociations for Indian audiences, was merely strange to most Americans. And Shantaram's painstaking direction, applied to material of this sort, produced a style of acting that was mannered by Western standards. Yet *Shakuntala* might, in a later day, have fared better.

A few months after *Shakuntala* went west the United States Supreme Court, in *U.S.* v. *Paramount et al.*,[5] handed down a decision that was to have a profound effect on the film world. As a result of this antitrust decision, the climax of years of litigation, the major American film companies were ordered to divest themselves of the ownership of thousands of theatres in the United States, and to desist from block-booking contracts with other such theatres. The companies involved in the decision were Fox, Loew's (including MGM), Paramount, RKO, Warner Brothers, Columbia, United Artists, Universal. Through ownership of key theatres and block booking, the court found, these companies had acquired a degree of control over the American market that had virtually closed it to other producers, except on terms dictated by these companies. The success achieved by the condemned practices may be suggested by the fact that in one year—1943–44—the eight companies had received over 94 percent of the total United States film rentals.[6]

No governmental quota had at any time blocked entry of foreign films into the United States. Yet the joint control over theatres by the eight major companies had created an effective private barrier. The art-theatre movement, struggling into life in spite of this, was still confined to a mere handful of theatres.

During the decade after *Shakuntala,* the eight major companies, no longer controlling their home market and also shaken by the rise of television, sharply curtailed their theatrical production. Theatres, no longer controlled by these producers, turned increasingly to foreign films. By the end of the 1950s a thousand American theatres would be regular outlets for foreign films. Westward journeys would now find a wider market waiting.

[5] 334 U.S. 131 (1948).

[6] Conant, *Antitrust in the Motion Picture Industry,* pp. 44–46.

Meanwhile in Bombay, formula was king. The formula, as dictated by exhibitor and distributor, called for one or two major stars, at least half a dozen songs, and a few dances. The story was of declining importance. It was conceived and developed toward one objective: exploitation of the idolized star. The subject matter, with increasing concentration, was romance. An overwhelming number of Bombay films now began with the chance acquaintance of hero and heroine, often in unconventional manner and novel setting. In backgrounds and characters there was strong bias toward the glamorous. Obstacles were usually provided by villainy or accident, not by social problems. Dance and song provided conventionalized substitutes for love-making and emotional crisis.

All this was to some extent in the ancient Indian tradition. In *Shakuntala,* too, we find the erotic balanced with decorum. *Shakuntala* also began with "boy meets girl," with the introduction arranged in novel style by a troublesome bee. In a summary of the content of Sanskrit plays, S. C. Bhatt tells us: "The plays know love at first sight, not arranged marriages."[7] The obstacles, as in the formula film, were provided by villainy or accident. And the humdrum background was shunned in favor of sylvan glade or palace.

But though the pattern was traditional, critics of the Bombay film preferred to consider it largely a Hollywood infiltration. In an India dominated by the arranged marriage, the impulse to do this was natural. Besides, Hollywood was working the same dramatic vein and was thus a ready target. In addition, many producers had begun a conscious introduction of Westernized details, for attention value. In matters of costume, Indian censors had at times allowed some degree of physical exposure in foreign films on the ground that customs differed. Indian producers always protested this "double standard" but meanwhile responded by introducing sequences in Western-type night clubs, such as cater to foreigners in Bombay and Calcutta. If an Indian hero was going through a period of disintegration he would no longer, in the old-fashioned manner of Devdas, visit an Indian "dancing girl" prostitute; he would go to a night club and watch high-kicking chorus girls while guzzling illicit whiskey.

[7] Bhatt, *Drama in Ancient India,* p. 67.

Jhanak Jhanak Payal Baje (JANGLE, JANGLE, SOUND THE BELLS)
SHANTARAM SPECTACLE, 1955. RAJKAMAL KALAMANDIR

(". . . drinking scenes . . . meant to condemn drink will not be cut out.") Some critics also felt that the traditional dacoit (robber) sequences were borrowing in technique from American gangster films. All this raised indignation. The theme of Hollywood influence, and the need to purge it, ran through the postwar years, even in statements of industry spokesmen.

In 1951, when the Russian director Pudovkin along with the actor Cherkasov visited India and received honors in various film centers, Chandulal Shah, as president of the Motion Picture Society of India, begged the Soviet visitors to defer their verdict on Indian films at least for a decade, "by which time the people would have lived down the effects of foreign rule and the influence of American films." The actor Cherkasov contributed to the spirit of the occasion by telling his hosts that Soviet actors were "never called upon to play the roles of gangsters, murderers and thieves, either in real life or on the screen."[8]

Meanwhile Bombay pursued relentlessly its formula of a star, six songs, three dances. Despite vocal critics, it had not the slightest reason to abandon the pattern. Throughout India, theatres were crowded. At low-price windows, lines waited.

The market for Hindi films was growing, and in this growth Hindi songs played a leading role. Here, too, the accusation of "Westernization" was increasingly heard. Most producers had by now adopted Western instruments and combinations of instruments. Many Bombay producers were using lush combinations of fifty or sixty instruments, foreign to Indian classical tradition, and film audiences responded with ecstasy. They had responded likewise to the borrowing of American jazz and Latin-American rhythms. These were transformed beyond recognition; to Western ears the result sounded Indian, but to Indian purists it sounded Western or "hybrid." "Hybrid music" became a favorite term of reproach, and the anger with which it was used mounted with the growing popularity of the film songs.

By the early 1950s the film song had become a key to successful film promotion. Exploitation had acquired a fixed pattern. Virtually all film songs were automatically made into phonograph records under the label "His Master's Voice"—an Indian subsidiary of a British company. Dubbed directly from the film sound tracks, the phonograph records were made during production of the film. Shortly before the premiere of the film, the songs would reach the air.

At first it was All India Radio that introduced film songs, but in 1952 this government-operated radio system began a drive against "hybrid music." At about the same time Ceylon was developing a commercial short-wave service beamed to India. It depended on advertisers of manufactured goods—many of them American—and needed a mass market. Radio Ceylon seized almost immediately on the Indian film song, disdained by All India Radio, as the key to its problem. Before long Radio Ceylon was devoting all peak evening hours to Indian film songs—mainly Hindi songs from Bombay. Soon Radio Ceylon dominated the air over India. By the mid-1950s

[8] *Journal of the Film Industry,* January, 1951.

young people in various parts of India were listening nightly to Hindi film songs on Radio Ceylon. The high point of the week became the Wednesday night Binaca Toothpaste Hour, a one-hour program on the Hit Parade formula. Sponsored by a toothpaste manufactured by the international Ciba company, each program was climaxed by the dramatic announcement, preceded by trumpet fanfares, of the leading song of the week. In many northern and central Indian cities, that moment found clusters of people huddled around tea shops or other places with radios. At the same time the *BMPA Journal* reported: "Beggars, the vociferous among them, are become proficient imitators of film songs, and their dividends largely depend on how best they are able to sing a film song."[9]

If all went well, at least one song in a film would be a hit before release of the film. The impact of this could be noted in the theatres. At the first strains of the already familiar song, an approving groan swept through the hall. In numerous instances films with hit songs, though disastrously reviewed, were box-office triumphs. Bombay saw no reason to change the formula.

Yet even in Bombay some producers, if only occasionally, swam against the tide. We shall menton them briefly.

One was K. A. Abbas. It will be recalled that he had written, partly based on his own experience, a 1941 film about a journalist under pressure from a business nabob. The success of this Bombay Talkies film, *Naya Sansar* (New World), had launched Abbas on a series of "socially significant" films. In 1949 one of these films, *Dharti Ke Lal* (Children of the Earth), became the first Indian film to be shown in Moscow.[10] It also ran in Paris and London. Telling a story of rural indebtedness and dispossessed peasantry, it used no professional actors. Abbas had been one of the founders of the Indian People's Theatre Association, dedicated to production of socially significant plays; *Dharti Ke Lal* had begun as an IPTA play and the film was produced under its auspices.

If Abbas scored a success with *Dharti Ke Lal,* he won a critical triumph with *Munna* (The Lost Child), produced by his own production company, Naya Sansar Productions (New World Produc-

[9] *BMPA Journal,* October, 1955. [10] *Journal of the Film Industry,* April, 1949.

tions)—named after his first film success. *Munna,* made in Hindi in 1954, was shown at the Edinburgh film festival in 1955 and at a festival of Indian films in Moscow a year later. It was singled out for praise by Prime Minister Nehru. It was considered by the Edinburgh *Scotsman* the "worthiest film of this year's festival,"[11] while the Manchester *Guardian* called it the festival's "most delightful surprise . . . it shines with gaiety and imagination."[12]

The leading character of *Munna* was a seven-year-old boy who escapes from an orphanage. Searching for his identity, he becomes involved in various segments of the world—and underworld—of Bombay.

Perhaps the most radical aspect of *Munna,* hardly noted by its Western admirers, was that it was the first Hindi film ever made without songs or dances.[13] Since the production of the first Indian sound film in 1931, it had taken twenty-three years for a producer to dare this experiment. Unhappily, while *Munna* received glowing praise, it was "not a box-office hit" in India.[14] It thus inadvertently bolstered the authority of those distributors looking for strict adherence to formula.

Even more ironic was the triumphant box-office success of another Abbas-written film which did follow formula. It was written by Abbas for a star who was rapidly becoming the most popular film figure in India—Raj Kapoor. Son of the distinguished actor Prithviraj Kapoor, who among his hundreds of roles had played Alexander the Great in *Sikander,* Raj Kapoor had started as clapper boy at Bombay Talkies. Soon getting a chance to act, he quickly became the idol of young India. Handsome and vigorous, he also had a talent for easygoing comedy, and for good measure was a fine player of the *tabla,* an Indian drum, a talent which was utilized in most Raj Kapoor films. In the postwar period he began production under his own banner, R. K. Films. Attracted by the proletarian themes of Abbas, he undertook such a theme—with a more sentimentalized

[11] *Scotsman,* August 30, 1955.
[12] *Manchester Guardian,* September 2, 1955.
[13] A Tamil film without songs or dances, *Antha Naal* (That Day), had also appeared in 1954.
[14] Holmes, *Orient,* p. 14, and interview, Abbas.

treatment—in *Awara* (The Vagabond). Abbas wrote the story and collaborated on the screen play. The film had its quota of songs, one of which, "Awara Hun" (I Am a Vagabond), swept through Asia. *Awara,* dubbed into Turkish, Persian, and Arabic, broke box-office records in the Middle East. Raj Kapoor and his feminine co-star, Nargis, became popular pin-ups in the bazaars of the Arab world. Meanwhile in Moscow, where Abbas had won a showing with *Dharti Ke Lal* without causing a furor, *Awara* swept all before it. The Soviet Union is said to have made a massive distribution of *Awara,* dubbed into a number of languages.[15] Prints were even flown to two Soviet expeditions near the North Pole. The Soviet distribution began in 1954, after Raj Kapoor, Nargis, Abbas, and others had visited Moscow as members of a film delegation. On a return visit to the USSR two years later, Raj Kapoor and Nargis were astonished to find themselves well-known film personalities. Bands played "Awara Hun" at airports. The climax of this visit was a puppet show performed by a leading Russian puppeteer. In the final play the chief puppets represented Raj Kapoor and Nargis.[16] The Russian enthusiasm led to immediate Soviet distribution of another Raj Kapoor song-and-dance film, again written by Abbas—*Shri 420* (Mr. 420.) It dealt with a character who specialized in fraud, that is, violations of Section 420 of the Indian penal code. It likewise had wide success in the USSR, though not on the same spectacular scale as *Awara.*[17]

Shri 420 was one of a number of Indian films purchased by the USSR in 1955, largely Hindi films from Bombay.[18] The price is said to have averaged Rs. 50,000 per film.[19] This bonanza caused considerable stir in India. According to some observers, a number of Bombay producers subsequently injected "proletarian" angles into films in production, in the hope of sharing in the next Soviet windfall. Almost all were disappointed. The Soviet Union had taken a

[15] Interview, Pochee. The Russian title was *Bradyaga* (Vagabond).
[16] *Filmfare,* December 21, 1956.
[17] The success of *Gospodin 420,* as the Russians called the dubbed *Shri 420,* was mentioned by John Gunther in *Inside Russia Today.* Assuming the film to be Russian, he erroneously described it as portraying a violator of Section 420 of the Soviet penal code. Gunther, *Inside Russia Today,* p. 325.
[18] *BMPA Journal,* July, 1955.
[19] Interview, Pochee.

RAJ KAPOOR IN *Shri 420* (MR. 420) OR *Gospodin 420,* 1955. R. K. FILMS

RAJ KAPOOR'S *Jagte Raho* (KEEP AWAKE), WINNER OF GRAND PRIZE AT KARLOVY VARY FESTIVAL, 1957. R. K. FILMS

RISE OF RAJ KAPOOR: *Awara* (THE VAGABOND), 1951, SPECTACULAR SUCCESS IN THE USSR. R. K. FILMS

fancy to Indian film songs and dances. The selections that followed were strictly formula. It could take care of its own propaganda.

Another producer-director who in the early 1950s was sometimes swimming against the tide was Bimal Roy. Born in Dacca, East Bengal—which eventually became Bangladesh—Bimal Roy had begun his career at New Theatres in Calcutta, and was cameraman for Barua's *Devdas, Mukti,* and other films. As the fortunes of New Theatres declined rapidly after the war, Bimal Roy moved to Bombay and began to make Hindi films. In 1952 he launched Bimal Roy Productions and the following year was acclaimed for his *Do Bigha Zamin* (Two Acres of Land), which won the Prix Internationale at the 1954 Cannes film festival and the Prize for Social Progress at the Karlovy Vary film festival.

Do Bigha Zamin, strongly influenced by the Italian neorealists, tells of the struggles of Sambhu, a Bengali peasant, to keep the two acres of land on which he and his family live. Years of drought have put them badly in debt. To save his land, Sambhu becomes a ricksha puller in Calcutta. There his son becomes involved with thieves. Sambhu somehow holds his family together and returns to his village with the needed money. But it is too late: the land has been lost and a factory built on it.

Bimal Roy showed himself to be one of the first Indian directors capable of simplicity and understatement. He showed these qualities even more surely in his later *Sujata* (Sujata). The girl Sujata is an untouchable. In her infancy she is sheltered in an emergency by an engineer and his wife, of Brahmin caste. The family assumes the arrangement to be temporary: a suitable home will be found for the child. But such a home is not immediately found. In the course of various postponements, the family becomes attached to the child and eventually she is brought up with their own daughter, of similar age. The untouchable girl is named Sujata—"well born." In later years Sujata is sometimes referred to by acquaintances of the family as "your daughter." The mother parries this but always says that Sujata is "like my own daughter." Sujata, though grow-

BALRAJ SAHNI IN *Do Bigha Zamin* (TWO ACRES OF LAND), 1953.
BIMAL ROY PRODUCTIONS

DOOMED HERO RETURNS: DILIP KUMAR IN *Devdas* (DEVDAS) TITLE ROLE, WITH VYJAYANTHIMALA, 1955. BIMAL ROY PRODUCTIONS

ing up in an atmosphere of love, is aware from this phrase that she is in some way different from the girl she regards as her sister.

Once they have accepted her presence, the foster parents are not especially troubled by Sujata's outcast origin until it is time to arrange the marriage of their own daughter, Rama. A marriage has been discussed for years with another Brahmin family, whose scion is the handsome and promising Adhir. This family is deeply disturbed that an untouchable girl has so long lived in Rama's home. However, they consider themselves modern and feel the marriage should go ahead, provided the untouchable girl is first married off to a suitable husband. They wish above all to avoid having her present at the wedding. Again there is an effort to make suitable disposition of Sujata. But a complication develops. Adhir, during his few visits to the house, has fallen in love with Sujata and wishes to marry her, not Rama.

If *Sujata* had been written and directed by Prince Barua, it is likely that this dilemma would have been sidestepped by one or

UNTOUCHABLE: NUTAN AS *Sujata* (SUJATA), WITH SUNIL DUTT, 1959.
BIMAL ROY PRODUCTIONS

more deaths. But Bimal Roy's *Sujata* asserts this marriage is possible. The young man takes a firm hand with his family; without persuading them, he has his way. Rama's mother is at first outraged that the long-planned marriage of her own daughter has been frustrated by the outcast girl they have sheltered, but in the end she relents and even calls her "my daughter."

The story is told by Bimal Roy in a series of deft, restrained episodes. They never lapse into the self-pity that mars many Indian chronicles of doomed heroes and heroines. Song is used, but most often as background effect.

Another Bombay film maker who sometimes appeared to stand out from the crowd was Guru Dutt, a producer-director-actor of unusual finesse. While he seemed to accept the Bombay pattern as a business necessity, he was obsessed with dissatisfaction over his own work, and in this respect seemed to be a spiritual descendant in Barua. In

GURU DUTT IN *Sautela Bhai* (STEPBROTHER).
COURTESY NATIONAL FILM ARCHIVE OF INDIA

Kaagaz Ke Phool (Paper Flower)—India's first wide-screen film, made in 1959—he played a Barua-like producer-director-actor who is making a new version of *Devdas,* the doomed-hero saga, and who eventually emulates Devdas by drinking himself to death. As in Barua's case, it later appeared that Guru Dutt had been rehearsing his own demise. For he, too, depressed over the failure of *Kaagaz Ke Phool,* went into a decline and apparently took his own life. After death his work acquired something of a cult following.

But K. A. Abbas, Bimal Roy, and Guru Dutt, chafing over formulas and occasionally rebelling, were by no means typical of the Bombay film world of the 1950s. There were few rebels. During this decade Professor Asit Baran Bose, sociologist at Lucknow University, did a content analysis of sixty Hindi feature films. He found the films dealt primarily with the unmarried and educated young of the upper and middle classes, living in cities. In roughly half the films the hero had no occupation, in almost two thirds of the films the heroine had no occupation. In most films the obstacles were provided not by a social problem but by an evil character. Most films had an evil male character, roughly half the films an evil female character. With an average of 7.7 songs per film, in 70 percent the hero sang, in 23.3 percent the heroine sang but did not dance, in 70 percent she sang and danced. In roughly half the films the hero lived alone; in one third he lived in a family. The heroine generally lived with a family. Rarely did hero or heroine live in a joint family.[20]

The young people whose love for each other was the main concern in these films moved through a diversity of settings, exuding vigor and radiant health and usually surrounded by consumer goods. Among Hollywood influences, the consumer goods seemed especially prominent. Always singing at the top of their voices—via the voices of "playback singers"—the young people went motoring, motorcycling, speedboating, skiing, waterskiing. Always the lavish background, radiant health, laughter. Seldom the joint family, the arranged marriage, work, and poverty.

Critics liked to maintain that the industry had foisted such diversion on the people, had "conditioned" them to it, so that now they knew no better than to want it. More likely, producers were exploiting drives that were very much a reality. Perhaps President Sukarno of Indonesia, when he addressed assembled film executives in Hollywood in 1956, threw some light on what it all meant. He hailed these executives, to their surprise and perhaps discomfiture, as fellow revolutionaries, and thanked them for their aid to the national revolutions of postwar Asia. By showing ordinary people with refrigerators and motor cars, he said, they had "helped to build up the sense of deprivation on man's birthright." Millions of people, he suggested to them, would never again be content to lack those things, and had acquired an irresistible determination to have them. "That is why I say you are revolutionaries, and that is why I salute you."[21]

[20] As reported in *Statesman,* December 12, 1959.
[21] Reported in *Variety,* June 6, 1956.

Was the formula of the Bombay film perhaps not as "escapist" as it seemed? Was it perhaps a step in a continuing revolution? Was it perhaps more closely related than many suspected to the five-year plans—whose essential base was an end of apathy and acceptance?

Perhaps underneath there was this meaning, but it was hardly in the minds of those who made and distributed films. They had found a pattern that worked. It was maintaining lines at Indian box offices. It was winning regular export markets for Hindi films in Africa, the Middle East, Burma, Malaya, Indo-China, Indonesia, and the Fiji Islands.[22] A few films went on into wider worlds. There was no reason to change.

Ordinary, decent, superdecent

In the Bombay of the 1950s the salaries of the healthy and beautiful stars kept climbing. But workers on the lower rungs knew a different story. In 1955 the government of Bombay State[23] decided to make an "Enquiry into the Conditions of Labour in the Cinema Industry in Bombay State." Made at the suggestion of the central government, the study involved interviews with 621 film workers and resulted in a report issued a few months later.

The enquiry found 25 studios in operation in Bombay in 1955. Eighteen of these were available for rent, and half of those were used exclusively for rental purposes. Of the innumerable production companies only one, Shantaram's Rajkamal Kalamandir, had its own studio *and* its own laboratory. Of the 25 studios, 11 had changed hands at least once since the war. There were 11 laboratories, of which 7 had changed hands at least once since the war.

The technical staffs maintained by studios and laboratories were, on the whole, better off than the much more numerous workers hired on a free-lance basis. But even in such staff categories as assistant cameraman, dollieman, assistant carpenter, electrician, and lightman, earnings averaged less than Rs. 100 per month.[24] Wage payments were often months in arrears.[25]

[22] *Screen Year Book, 1956,* p. 229.
[23] Later divided into two states, Maharashtra and Gujarat.
[24] *Report on an Enquiry into the Conditions of Labour,* p. 48.
[25] *Ibid.,* p. 51.

It was among free-lance workers that conditions were most harsh and chaotic. The report states bluntly that since 1947 "the industry has been disorganized and polluted"[26] and its affairs handled "by men at the top in an unscrupulous way."[27] Late payment and non-payment of wages were common. The report stated:

> It appears that it has become a regular feature in the Film Industry to change the name and label of the concern in order to get rid of the claims of the employees and, if possible, of all the creditors.[28]

A union told the enquiry that during the postwar decade the film workers of Bombay had lost more than Rs. 10,000,000 in unpaid wages.[29]

Written contracts were almost unknown. The enquiry also revealed:

> In a large number of cases the employers obtained the workers' signatures for receipt of wages on blank vouchers. The date and amount is entered on the blank vouchers according to the requirements of the producers. This is one of the methods of adjusting "black money."[30]

The enquiry found 62.3 percent of free-lance workers to be "under-employed."[31] But in spite of the "unsatisfactory conditions," it found that

> the workers hardly think of leaving the industry. . . . There are many instances of "light boys" and peons becoming directors and producers. These instances serve as a pole star to the film employees and they remain attached to the industry.[32]

While the report threw light on various kinds of workers, it gave particular attention to "junior artistes" or extras, and especially to women in this group. The report thus provides an interim glimpse of the status of women in the film field. It will be recalled that in the 1910s even prostitutes shied away from film but had gradually overcome their reluctance. In the 1920s the Indian Cinematograph Committee had concerned itself with the problem of securing women "of the better classes" for the film field. During this decade the situation had been somewhat improved by an influx of Anglo-Indian girls. These casteless children of mixed marriage,

[26] *Ibid.*, p. 7. [27] *Ibid.*, p. 11. [28] *Ibid.*, p. 38. [29] *Ibid.*, p. 37.
[30] *Ibid.*, *p. 39.* [31] *Ibid.*, p. 19. [32] *Ibid.*, p. 11.

never fully accepted by Indian or British circles, had found a welcome in film, and this had resulted in such silent film favorites as Sita Devi (Renee Smith), Sulochana (Ruby Meyers), Indira Devi (Effie Hippolet), Lalita Devi (Bonnie Bird), Madhuri (Beryl Claessen), Manorama (Winnie Stewart), and Sabita Devi (Iris Gasper).[33] A number of these had dropped out in the sound era because of language problems, but some had continued.

In the 1930s the appearance on the screen of such women as Durga Khote and Devika Rani, both of Brahmin caste, began a more radical transformation in the popular image of the screen actress. The change was irregular but nonetheless marked. The role of Devika Rani was especially influential in that she was married to a highly respected film leader, was partner in a leading production company, and eventually became its controller of production. In a cohesive concern like Bombay Talkies, where the management watched over virtually all aspects of the lives of the company, banned drinking, and concerned itself with the educational development of the staff, the qualities of a Devika Rani could not fail to have an impact. Here and there "respectable families" began to allow their daughters to interest themselves in screen careers.

Devika Rani retired from Bombay Talkies in 1945 and did not again appear on the screen.[34] There were changes in the film field she had not liked. The kind of company Bombay Talkies had represented disappeared in the postwar years, and Bombay Talkies itself died a lingering death. Its last film appeared in 1952; thereafter its premises served as a rental studio.

During those postwar years screen stars continued to hold prestige. Such rising stars as Nargis and Vyjayanthimala appeared often on platforms with statesmen. But at other levels there were changes. The "junior artiste," no longer a member of a cohesive company, was now part of the floating population of "underemployed" freelance workers. The *Report on an Enquiry into the Conditions of Labour* revealed that most hiring of "junior artistes" was, by 1955, done through "extra suppliers." These agents received 10–25 percent commission from the "artiste."

[33] *Who Is Who in Indian Filmland,* pp. 47–74.

[34] In the same year she married the Russian artist Svetoslav Roerich.

In placing an order for extras, the producer customarily specified one of various grades. He told the supplier he wanted an "ordinary girl," a "decent girl"—this category was subdivided into classes A, B, and C—or a "superdecent girl." Rates for the categories were standardized. A superdecent girl was one who would seem acceptable in a high society or court setting, whereas an ordinary girl might appear in a crowd scene.

The system of ordering extras was said to lead to abuses. The report stated that the extra suppliers "have little social and cultural background," and pictured them as rapacious:

> When an order is received from the producer for a "decent A class girl," the agent supplies that category of artiste and receives payment accordingly from the producer, but while paying wages to the artiste he usually pays her the wages of a "B class girl" on the plea that the producer required an artiste of the latter category and that he had done her a favour by securing her the day's work. The difference in wages is pocketed by the agent. The artiste is helpless, as she has no access to the producer.[35]

The wage scale for a day's work by a "junior artiste," as paid by producers, was given as follows:[36]

Classification	*Daily Wage*	*Commission deducted*
Ordinary girl	Rs. 5	10%
Decent, Class C	10	20
Decent, Class B	15	25
Decent, Class A	20	25
Superdecent	25–40	25

Total earnings of ordinary girls averaged Rs. 17 per month; of decent class C girls, Rs. 33 per month; of decent class B girls, Rs. 54 per month; of decent class A girls, Rs. 120 per month; of superdecent girls, Rs. 175 per month. Those who could dance averaged Rs. 194 per month.[37]

The report offered statistics on "age of entry." Among free-lance employees more than 10 percent started their careers at an age below 14 years. Most began between 21 and 25.[38] Newcomers, said the report, "are prepared to forego their wages for the sake of a mere

[35] *Report on an Enquiry into the Conditions of Labour,* p. 21.
[36] *Ibid.,* p. 25. [37] *Ibid.,* p. 36. [38] *Ibid.,* p. 21.

appearance on the screen." The wages in such cases were "swallowed by the agent."[39]

The system of recruiting extras through middlemen, said the report, "breeds immoral practices. . . . Under threat of unemployment and starvation, the artistes cannot but succumb to the dictates of the supplier."[40]

Unionization was found to be in a rudimentary state. No film union had existed before 1946. A union of studio and laboratory staff workers was formed that year, and had survived and grown. Because the film industry was spread over an area of twenty miles, most workers found it difficult to attend meetings. Moreover:

> Several workers stated that they were unable to pay even the small subscription of the union as they were not getting their wages on time. The financial position of the majority of unions is, therefore, far from satisfactory. In strange contrast to this, the workers getting fairly good salaries are not much inclined to join the trade unions.[41]

A Cine Writers Association and a Playback Singers Association had been launched but had ceased activity because of lack of support.

In its final chapter the report estimated: "The payments to the stars cost the producers anything from 31 to 50 percent of the entire cost of producing the picture."[42]

The enquiry was intended to include a study of conditions in Kolhapur and Poona, but it was found that production had virtually ceased in those centers. The final film of the Prabhat Company, in the Marathi language, had appeared in 1953.

Pageants for our peasants

Two phenomena marked the postwar decade in the Madras film world. They seemed to move in opposite directions.

One was the rise of southern linguistic nationalism. It was anti-Hindi, anti-north, and extolled the glories of the ancient Dravidian languages and culture. It made Hindi a symbol of a northern domination to be feared and averted. It became a highly emotional force in politics. It also became strong in the southern film world

[39] *Ibid.* [40] *Ibid.*, p. 22. [41] *Ibid.*, p. 73. [42] *Ibid.*, p. 79.

and made extraordinarily successful use of film in its drive for power.

The other phenomenon, ironically, was the successful entry of Madras into production in Hindi and its triumphant invasion of northern Hindi markets.

The man who engineered the push into Hindi areas was S. Srinivasan—usually known by the simplified version of his name, S. S. Vasan. He was an ebullient, versatile businessman whose many interests included a weekly journal in the Tamil language, of large circulation, and race horses. In 1938 he became involved in film by taking over distribution of films of the Madras United Artistes Corporation. In 1941 there was a fire in the studio of the Motion Picture Producers Combine, already briefly mentioned as a joint enterprise of various Madras producers. Like most studios in India, it was uninsured, because no insurance company would take the risk. The partners, by now at odds with each other, decided to sell the charred premises. Vasan bought them, did some rebuilding, and launched the production company Gemini Studios.

During the war period the new firm dabbled in a variety of films including a mythological, a stunt film full of magic, and a romance or two—in which Gemini, according to observers, injected a more sensual note than most south Indian producers had dared or cared to venture. Meanwhile Vasan was preparing his big postindependence move.

According to his own account, every part of the strategy was carefully planned. What the nation needed, after having "put up with many insipid war propaganda pictures,"[43] was pageantry. The result was *Chandralekha* (Chandralekha).[44] Representing the unprecedented investment of Rs. 3,000,000, it was released in 1948—first in Tamil, then in Hindi. No previous Madras venture into Hindi production had succeeded; there had been only two such attempts in over a decade of Madras sound production.

The Hindi release was prepared with particular care. Gemini had recruited two of the founders of Vauhini Pictures, who had pre-

[43] Vasan, "Pageants for Our Peasants," in *Indian Talkie, 1931–56,* p. 26.

[44] A print of *Chandralekha* has been preserved by George Eastman House, Rochester, N.Y.

MADRAS IN ACTION: VASAN'S *Chandralekha* (CHANDRALEKHA), 1948. GEMINI

viously seen service at Prabhat. One of these, A. K. Sekar, shared Vasan's taste for spectacle. For *Chandralekha* he designed not only the mammoth sets but also the mammoth promotion campaign. Vasan spared no expense. He was a believer in full-page newspaper advertisements—the front page, if available. In this way Gemini literally bludgeoned its way into northern markets. The tactics proved an unqualified success. Within India *Chandralekha* grossed Rs. 10,000,000.[45]

One of the highlights of *Chandralekha* was a great drum dance which seemed to cover acres of ground. Every girl in a huge ensemble danced on top of a gigantic drum. Other factors contributing to the box-office sweep of *Chandralekha* were daring horsemanship and trapeze work, dazzling swordplay, and, in the words of a Madras

[45] *Film Seminar Report*, p. 194.

MADRAS IN ACTION: MEIYAPPAN'S *Demon Land,* 1948.
A.V.M. PRODUCTIONS

film chronicle, "one significant mid-close shot of the heroine in lustful proportions."[46]

No Indian producer had ever offered such spectacle. Vasan at once went on to other pageantry, including a sequence in which a herd of elephants, in response to prayers to Ganesa the elephant god, thundered of their own accord across a landscape to the rescue of an imprisoned prince and, pushing in unison against the fortress walls, brought them toppling to the ground.[47]

Meanwhile another Madras company, A.V.M. Productions—founded by A. V. Meiyappan, who had started as a distributor of phonograph records—made similar assaults on the Hindi market. To clinch these markets, Madras producers now began to outbid Bombay for its leading stars. Indian Airlines Corporation began to carry a lot of film stars, music directors, and their families and reti-

[46] *Flash*. Silver Jubilee Issue, April, 1961.
[47] Made for the Tamil film *Avvaiyar,* the sequence was reused by Vasan for his Hindi film *Bahut Din Huwe,* although the plots of the two films were otherwise unrelated.

nue between Bombay and Madras, and the price of their services soared higher than ever. Pakshiraja Studio in Coimbatore applied the same technique; in 1955 the great Dilip Kumar and Meena Kumari, both approaching top starring rank, flew south to make *Azaad* (Azaad), the story of a "mountain bandit." The film grossed Rs. 8,000,000 in its Hindi version, and additional millions in other versions.[48] Southern producers were out-Bombaying Bombay. The result was no artistic renaissance. But Gemini had laid the basis for a shift. In share of Indian production, the post-*Chandralekha* era showed:[49]

	Bombay area	*Calcutta area*	*Madras area*
1949	60%	29%	11%
1954	50	19	31
1959	38	16	46

In Madras, even more than in Bombay, experiments were few and far between. Mention must be made of *Kalpana* (Imagination), which appeared in 1948, the year of *Chandralekha*. An independent production by the celebrated dancer Uday Shankar, who rented Gemini Studios for the purpose, it told its entire story in dance. *Kalpana* caused a stir in many parts of India, running for twenty-six weeks in Calcutta. It was felt by some admirers of the film that Uday Shankar had magnificently harnessed the powers of film and dance to the purposes of a banal story of "mawkish sentimentality." He was praised for his grasp of the film medium, and it was widely felt that he had opened a door to new vistas. But the experiment was so ambitious and specialized that few would be likely to follow in his footsteps.[50]

Of less distinction was the 1954 experiment of A.V.M. Productions, *Antha Naal* (That Day), the first of the songless, danceless features. Patterned after the Japanese *Rashomon,* it lacked the finesse of the original. Again, its commercial failure was an augury favoring the established patterns.

As in other centers, auguries of various sorts were highly respected in Madras. Producers consulted not only distribution reports but

[48] *Azaad,* p. 3. [49] Jain, *Economic Aspects,* p. 52.
[50] See the discussion by Chidananda Das Gupta, "Reminiscence," *Seminar,* films issue, May, 1960.

also astrologers. A new production was generally launched on a day and at a time set with astrological help. The industry was thus star-dominated in more ways than one. Productions were often started with appropriate observances, serving religion and personnel relations simultaneously. Trade papers often carried such items as: "The camera was switched on at 10:09 A.M. by the star's mother."

By 1955 star salaries, in the bidding against Bombay, were rumored to have reached Rs. 400,000 per film, in white and "black." Meanwhile in southern theatres a chief projectionist received a basic salary of Rs. 100–130 per month in the bigger theatres, Rs. 80–100 in the lesser theatres, plus cost-of-living increments. A chief ticket taker received exactly half as much. These scales had been set in a labor arbitration in 1947, and remained in effect.[51] Studio and laboratory employees lived on a comparable scale. As in Bombay, most free-lance film workers existed at more precarious levels.

"O divine Tamil"

While the south was pushing into northern markets, it was also the setting of a highly dedicated movement which was at first not taken seriously by state or national leaders, but which was destined to play a role of increasing importance. This was the Dravidian movement. The emotions involved were complex and had deep roots.

In southern India, alongside the national independence campaign, there had been for some decades another drive—the anti-Brahmin movement. The Brahmins, heading the Indian caste structure, were identified in the southern mind with the Aryans who, long before the Christian era, had started pushing down from the north and had imposed their rule on the older Dravidian civilization. Hindu culture had emerged from the synthesis of these two civilizations. In the oversimplified view favored by many non-Brahmins, the Brahmin caste was a final precipitate of the conquering Aryans, while the lower castes represented the old Dravidian order. This view gave the non-Brahmin castes a sense of solidarity in relation to the Brahmins.

[51] *In the Court of the Industrial Tribunal,* pp. 54–60. The pay scales rose sharply in the 1960s and 1970s.

During much of the British era, government positions and special privileges went almost exclusively to Brahmins. In a sense, the early policy of the British was to accept and adopt the caste structure, ruling through its top layer. It was not until the 1930s that the Brahmin supremacy began to be weakened by the non-Brahmin alliance.

The Brahmins had always fostered the study of Sanskrit, the Aryan linguistic heritage. Brahmins in Madras tended to look down on vernacular Tamil and to embellish their own Tamil with words coined from Sanskrit. In response, it became a matter of sacred duty among non-Brahmins to purge their speech of all Sanskrit taint. The anti-Sanskrit purge reached the height of its fervor around 1930.

During the 1930s, in anticipation of independence, there began a drive by many leaders of the National Congress to build Hindi—based on Sanskrit—as a future national language. It gathered wide support in the north but growing resistance in the south. The Dravidians looked on the imposition of any strange tongue as a badge of servitude. They had felt this way about the Sanskrit of the Aryan and the English of the British raj; they felt the same way about the Hindi of the north.

Under the reorganization promulgated by the British in 1935, the Brahmin scholar C. Rajagopalachari became chief minister of Madras in 1938. He was a brilliant, sharp-witted writer and speaker, a Sanskrit pundit, a friend of Gandhi and Nehru, a respected leader of the National Congress. He was also a man of strongly held convictions. One of his first acts was to decree compulsory study of Hindi in the schools of the province of Madras.

The move caused such a tumult, such firm resistance, that it was abandoned in two years. But after independence the need for a national language was again widely discussed. It seemed unthinkable that affairs of national departments, legislatures, and courts should continue to be conducted in English, a language understood by only a thin stratum of Indian society, and identified with colonial rule. In the view of many leaders, Hindi based on Sanskrit was the only conceivable choice. All India Radio began to feature a Sanskritized Hindi. The Constitution of 1950 specifically designated Hindi as the national medium, but the time of its official inauguration was postponed. Meanwhile, there would be time for education and evaluation of progress made. Most states adopted the study of Hindi, but the south had its own formula. Madras State—later Tamilnadu—required all children to study Tamil *and* one other Indian language, which might be Hindi. But as the anti-Hindi campaign mounted, the formal study of Hindi virtually ceased.

Every move toward the imposition of Hindi seemed to stiffen southern resistance. The issue provided the emotional base for various new associations and parties. One of these, the Dravida Munnetra Kazhagam or D.M.K.—the Dravidian Forward Movement—was formed in 1949. Its founder was a dramatist and sometime actor, C. N. Annadurai, who held a master's degree from the University of Madras and quickly won fame for the vivid imagery of his speech and his flair for impromptu alliteration in "chaste Tamil." Annadurai was, at about this time, becoming involved in film, and before long various film stars and writers rallied to the D.M.K.

In the early 1950s they began the casual introduction into films of symbols of their movement. References to the colors black and red, adopted as party colors, became frequent. The word *anna,* big brother—popular name for Annadurai, the party leader—was often used. Later the motif of a rising sun, adopted as party emblem, began to appear. Any such symbol would evoke wild applause in the theatre.

All this quickly developed into a popular game between audience and producer. It led to such dialogue as:

Man 1: The night is dark.
Man 2: Don't worry! The rising sun will soon bring light and good fortune.
(*Audience: wild cheers and applause*)

Or:

He: Believe me, sister!
She: I do, Anna, I do! The whole land believes in you, and will follow you.
(*Audience: wild cheers and applause*)

In the casual selection of a sari:

She: I always like a black sari with a red border.
(*Audience: wild cheers and applause*)

Two people lost in a forest:

Man 1: Should we turn north?
Man 2: No, never! South is much better.
(*Audience: wild cheers and applause*)

This was, of course, the very technique used by the Congress in earlier years. Again it proved its effectiveness. So noisy was the approval that greeted these casual Dravidian injections that producers unconnected with the movement began to use the symbols. They too wanted the applause. Actors found it prudent to associate themselves with the movement.

All these symbols, like the Congress symbols, appeared in stories that had no specific relation to the movement. Yet as a result of the symbols, the stories occasionally seemed to take on new meanings. In the mid-1950s M. G. Ramachandran, one of the two most popular of south Indian stars, began to specialize in stunt roles in the Douglas Fairbanks tradition. He appeared as folk-hero, battling royal usurpers and their henchmen, fighting against incredible odds. Many critics considered the stories purely "escapist." Ramachandran expressed the opinion that they were not escapist. To his public—so Ramachandran explained—the long-entrenched Congress leadership in New Delhi had become a species of royalty, and the folk-hero symbolized the southern Dravidian, struggling against odds to establish justice.

One such folk-hero film, *Nadodi Mannaṇ* (The Vagabond King), was vaguely set in an earlier era. Its opening song began: "O divine Tamil, we bow to you, who reflect the glories of ancient Dravidians."[52]

Throughout this period, film actors and writers played a role in party rallies. These rallies were usually advertised as occasions for meeting stars. One of the stars who appeared in such rallies, though not otherwise identified with the movement, was N. S. Krishnan.

[52] From the film *Nadodi Mannan,* produced by Emgeeyar Pictures.

MADRAS IN ACTION: RAMACHANDRAN'S *Nadodi Mannan* (THE VAGABOND KING), 1958. EMGEEYAR PICTURES

A brilliant comedian and satirist, he was often compared with Chaplin. His following was confined to the Tamil-speaking area, but the intensity of its devotion appears to have matched anything in the annals of comedy. When he died in 1957, the crowds at his funeral procession are said to have been comparable to those at the funeral of Mahatma Gandhi.

Krishnan had built his huge following during the war years, when it had become almost obligatory for a Tamil film to have a Krishnan sequence, though unrelated to the main plot. Krishnan had always written such sequences himself, and acted them with his wife, Mathuram, and other associates. During the war period over a hundred such sequences were made by Krishnan for many different producers. By the beginning of the independence era he was a figure of towering stature.[53]

[53] A trial on a murder charge, of which he was acquitted, had the effect of increasing Krishnan's popularity. His attorney in the case was the statesman and man of letters, K. M. Munshi, who conferred with Krishnan several times "in the lockup" and later wrote of these meetings: "Even under the Damoclean

In 1948 Krishnan produced a feature film which Annadurai wrote. As the party began its organization, Krishnan appeared often at its rallies and gave enormous impetus to its growth. The skits he performed on such occasions often satirized orthodoxy. In some there was ridicule of Brahmins. Such features were always sure of an ovation at D.M.K. rallies.

The general council of the D.M.K. party seldom met, and when it did, seldom discussed its platform or principles. The party stood for southern autonomy, but the details of such policy were seldom debated, either in council or executive committee. One of the early D.M.K. leaders, E. V. K. Sampath, left the party because of this. He protested that the business discussed at council or committee meetings was usually the dispatching of stars to this or that rally.[54] To its detractors the D.M.K. was a fan club. It was that, of course, but it was a new kind of fan club. No other fan club had given its members this sense of involvement in national affairs and great causes.

To the actors, too, this was something new. The star M. G. Ramachandran expressed his feelings by saying that Douglas Fairbanks, great as he had been, was already forgotten. An acting career must now, Ramachandran felt, have an added dimension if it was to have real meaning. He had found that dimension in the Dravidian movement.[55]

While appealing to long-submerged feelings, the D.M.K. also had practical appeals. Schoolteachers were urged to see the imminence of official Hindi as a threat to their livelihood and influence. Young people saw it as an obstacle to political careers. Already, according to southerners, New Delhi was giving preference in civil service to those who could speak and write Hindi. Such a policy could only lead to dominance of northerners in the central bureaucracy. There were also cultural arguments. The English language, it was now

---

sword which was hanging over him he would cheerfully mimic Mr. Justice Mockett, myself and my learned opponent the Advocate-General—and all with such inimitable grostesqueness as to throw me into fits of irrepressible laughter. Never for a moment had I thought of myself as cutting so ludicrous a figure in a court of law." Quoted in *Indian Talkie, 1931–56,* p. 136.

[54] Interview, Sampath.

[55] Interview, Ramachandran.

said, had at least opened the door to new worlds of knowledge and ideas. Hindi, the Esperanto of the north, would open no such doors.

In the independence era, C. Rajagopalachari returned to state office. Representing the Congress party, he was chief minister of Madras State from 1952 to 1955. During these years it was clear that the D.M.K. was building strength, but most government leaders regarded it with amusement. Rajagopalachari hardly noticed it. K. Kamaraj, who succeeded him as chief minister in 1955, again representing the Congress, was openly scornful. He was quoted as asking: "How can there be government by actors?" For "actors" he used the most contemptuous term available—*koothadi,* or mountebank. The contempt rallied the screen world to the D.M.K.

The time would come only a few years later when this troupe of mountebanks would stagger the Congress at the polls and emerge as the dominant force in the politics of Madras State—which would become Tamilnadu. The Dravidian movement would experience complex schisms and alliances, but it would meanwhile propel into political careers a number of film personalities, some of whom would become chief minister of the state. These would include C. N. Annadurai, big brother, whose political career would be cut short by death, but whose name would remain a party symbol; and later M. G. Ramachandran, whose screen image as champion of the oppressed would so merge with his political persona as to overwhelm opposition.

But all this was inconceivable in the early 1950s, when the D.M.K. was just beginning its career of booking stars for quasi-political rallies of movie fans. At that time the D.M.K. was largely an oddity. The big film story then was the rise of Gemini Studios and A.V.M. Productions in the Hindi market. Both were building national distribution organizations, the only Indian producers so organized, and were beginning to have a look of national stature.

S. S. Vasan of Gemini, at the crest of his triumphs with *Chandralekha* and later pictures, was rapidly becoming an acknowledged leader of the all-India film industry. A constant traveler, he crusaded with characteristic vigor against the heavy taxes imposed on the film industry. In 1953 he arranged a large meeting of film celebrities in Madras. To this he invited C. Rajagopalachari, the chief minister of the state.

C. Rajagopalachari had long been outspoken in his dislike for films. He often advised people to avoid them. He felt the Indian industry was imitating the producers of "the smutty west," who, in his opinion, "create sin." He felt that the view of life constantly shown on the screen was adding to "man's quantum of sex urge," and that this was precisely what India did not need.[56] But he accepted the invitation from Vasan.

As he sat quietly on the platform, speeches began. One was by Vasan. He said that some people always spoke of the film industry as some sort of educational institution. It was nothing of the sort, he suggested. A film, to meet its cost, had to please 10 million people. If films were made to please social reformers, they would not even reach 15,000 people. The real function of film, said Vasan, was to entertain, and to that end the producer should dedicate himself. Next he spoke of taxes. He mentioned *Chandralekha* and its domestic gross of Rs. 10,000,000, at least half of which had gone into taxes. He gave an estimate of what the film industry as a whole was contributing to central, state, and local exchequers. He insisted it was an excessive burden and that parts of the film world were being bled to death. He suggested the governments should mend their ways. Otherwise, he proposed, it would be a sound idea for the film industry to stop production for a while and halt the ample flow of tax revenue.

C. Rajagopalachari was invited to comment if he wished to. He did wish to. He welcomed the fact that Mr. Vasan had disclaimed an educational role. Education, he agreed, required qualified people. As to taxes, he reminded Mr. Vasan that he, Rajagopalachari, had campaigned for prohibition, and had thus worked to put an end to another corrupting influence, even though it had brought in more taxes than the film industry. He urged Mr. Vasan not to pursue this argument. However, Mr. Vasan had suggested that the film industry should voluntarily stop producing. If the industry itself should actually devote its attention to "reducing this poison," and terminate itself, that would be something unique, and an extraordinary service. If it should end entirely, he would not be sorry,

[56] *Indian Express,* August 6, 1953.

the minister said. He assured Mr. Vasan of his support.[57] The *Indian Express* reported further: "The function concluded with a dance performance by Vyjayanthimala."[58]

A 40 percent loss

While Bombay was holding its lead as the major film center and while Madras, spurred by linguistic nationalism, was consolidating its strength in the south and invading northern Hindi markets, the film world of Calcutta was in the doldrums.

No other production center had been so injured by partition. Of the Bengali-language market that had been its domain, 40 percent had become part of East Pakistan. Part of the Hindi-language market had also become foreign territory, in West Pakistan, but this was a minor loss to Hindi producers compared to the truncation suffered by Bengali producers.

The failure of India and Pakistan to agree on an exchange rate for their respective rupees soon proved costly to Calcutta producers. Pakistan decreed its own exchange rates, with the result that film rental remissions fell sharply. There were also reports of Pakistan levying a special "penalty" on Indian films, although the nature of this penalty was not clear and reports were contradictory.[59] Meanwhile the Indian government added to the problems of film exporters by its "reimport" policies. When prints came back from East Pakistan, the only Bengali-language market outside India, they were treated by India as new imports of "exposed film" and an import duty was levied. This duty had been designed originally as a way of deriving revenue from importers of foreign films. Now this import duty, exceeding the laboratory value of the print, fell heavily on the Calcutta film world. The economics of export to this limited area became forbidding. At the same time the small Pakistan film industry was beginning to ask for protection from Indian competition, and eventual moves in this direction were considered certain.

[57] Account based on *Hindu,* August 6, 1953; *Indian Express,* August 6, 1953; and interviews with film makers.

[58] *Indian Express,* August 6, 1953. Years later Rajagopalachari became an anti-Hindi leader.

[59] *Report of the Film Enquiry Committee,* pp. 120–21.

All these problems were accompanied by increases in production costs and the sharp rise in taxes. Throughout the postwar decade this combination of blows cast gloom over the Calcutta film world. Editorials in the *BMPA Journal* sounded, month after month, like speeches of a doomed film hero.

> GATHERING CLOUDS
>
> The depression is already here. Clouds are gathering in the sky over Bengal's film-land. They are dark and laden with ruinous rains. All in the film-land should congregate, to confer, how best to ward off the coming calamity. The strong should now help the weak to get a foothold and the weak thus helped, should remember to reciprocate when today's strong become weak later.[60]

Ironically there was, throughout this period, a surface prosperity for Bengali films. As elsewhere in India, lines waited at low-price windows; tickets for hits disappeared into the black market. This was partly because the central government, after 1948, had called a virtual halt to cinema building, to save raw materials. For years the growth of exhibition facilities had lagged behind the growth of production. A shrinking market now supported a bulging industry.

In 1951 M. D. Chatterjee, president of the Bengal Motion Picture Association, referring to the "queues in front of the theatres," said emphatically:

> This profile of prosperity is a deceptive facade. . . . Of the forty-three films in the Bengali language released in 1950 hardly a dozen will cover their cost and bring some return to the producers. . . . near-paralysis has afflicted the production industry in Bengal.[61]

For the year 1950, he estimated the loss on unsuccessful pictures at Rs. 2,500,000. He estimated an additional loss of Rs. 3,500,000 on pictures abandoned in the midst of production.

Not surprisingly, the years brought a steady drift from Calcutta to other centers, especially Bombay. Bimal Roy, Nitin Bose, Kidar Sharma, and others joined the Bombay migration. Barua, who had resisted the call of that "bazaar," died in 1951.

Since the beginning of the talking film era, Calcutta had not only supplied the needs of the Bengali-language market; it had also

[60] *BMPA Journal,* May, 1949.
[61] *Ibid.,* June, 1951.

taken a large share of the revenue from Hindi markets. It had obtained this share easily. It had simply produced films to its own taste, making Hindi versions along with Bengali versions, and watched the revenue come in. The profits from the Barua *Devdas* had come from many parts of India.

But in the postwar years Calcutta double versions ran into trouble. Between 1948 and 1955 New Theatres produced a number of double versions that succeeded in Bengali, but not in their Hindi versions. The Hindi titles were *Anjangarh* (Anjangarh), 1948; *Manzoor* (Let It Be), 1949; *Naya Safar* (New Venture), 1952; *Bakul* (Bakul), 1955.[62] Other producers were having similar experiences. For New Theatres, the setbacks virtually marked the end of the road.

Some said all this had happened because New Theatres, along with other Calcutta companies, had lost many talented men to Bombay. Others ascribed it to a change of tastes in the Hindi market.[63] Gemini Studios of Madras, with drum dances and thundering elephants, had caught the national fancy and brought forth a rash of epics. The quieter offerings of Calcutta were falling by the wayside.

To some it seemed that Calcutta would have to adjust itself to the new era and learn to produce "pageants for our peasants." Perhaps, like Madras, it would have to outbid Bombay for its stars, gambling for the highest stakes. Some moved in this direction, with minimal success. Calcutta simply did not have the *Chandralekha* touch. But it turned out there was another possible direction. Calcutta had always been a city of international awareness. It was the city of Tagore, citizen of the world. And while some in Calcutta film circles saw as their only hope the Hindi-language markets of India, there were others who began to think in wider terms.

To describe the genesis of this thinking we must turn aside for a moment from the commercial film world of Calcutta to something that had struggled into precarious life at the moment of independence. It seemed for years to have no chance of surviving, yet something kept it stirring. It is important to our story not for what

[62] Interview, Pramanick.
[63] Interview, Sircar.

it was—a film society—but for the role it played in the education of several people.

Film society

The Calcutta Film Society was founded in 1947, the year of independence, by Chidananda Das Gupta and Satyajit Ray. Both were advertising men. Both worked in the Calcutta branch of the London advertising agency D. J. Keymer. Both were hungry to see and study films other than those appearing in Calcutta theatres, and the Calcutta Film Society was founded for this purpose. It was not the first such society—a Bombay Film Society had been formed in 1942.

In India film societies worked from the start under crushing difficulties. In the United Kingdom the law exempted nonprofit membership film societies from censorship regulations. In the United States the few states that had adopted censorship systems likewise exempted nonprofit membership organizations. Without these exemptions the National Film Theatre in the United Kingdom, Cinema 16 and the Museum of Modern Art film showings in the United States—not to mention film societies on many university campuses—could hardly have existed.

The Indian Cinematograph Act of 1918 and later amendments offered no such exemptions. Independent India retained the rigid restrictions of colonial India. No film society, university, school, museum, club, or business concern could show a film without submitting it for censorship.[64] The procedures were the same as for commercial theatre showings. After January, 1951, they included a fee of Rs. 40 per reel for films over 2,000 feet. A complete script, a synopsis in eight copies, and texts of songs in eight copies were needed.

[64] For a meeting of the Indian Academy of Sciences in Poona in 1949, a scientific film shot in the Arctic was flown to India. Held up for censorship, it reached the Academy meeting just in time for the scheduled screening. When the slide showing the censor's approval appeared on the screen, "Dr. Raman led the chorus of satiric approbation by lusty clapping of hands." *Journal of the Film Industry,* January, 1950.

For many of the experimental films shown at British and American film societies, no such material existed.

At the time of independence, Indian law exempted from censorship one kind of projection: a showing on diplomatic premises. This was to play an important role in the story of Indian film societies. In June, 1952, a second exemption was added. The rules now authorized a producer, without prior censorship, to project footage from his own unfinished film. Government representatives had to be admitted to any such screening at any time.[65] The new exemption did not help film societies.

Besides being subject to censorship, film societies had to pay entertainment taxes. This was another requirement from which British and American film societies have been exempt. Entertainment taxes continued to be levied on the Calcutta Film Society until 1960, when officials recognized it as an educational organization. But a new exemption application had to be made for each showing.[66]

Still another hurdle facing Indian film societies has been import duties. Foreign producers were sometimes willing to sell prints at low cost for nonprofit use in India, but the government took the position that import duty had to be paid. It long resisted the notion of classifying any feature films as educational material.

In spite of all these difficulties the Calcutta Film Society began in 1947 to hold regular showings of films of special interest as works of art. Membership rose steadily.

The society found it had, essentially, three sources of material. The first was the Central Film Library of the Ministry of Education. Formed as a service to schools, it had bought a few film classics such as *Nanook of the North*. But the emphasis in its purchases was on classroom teaching films. The Calcutta Film Society found it could obtain from the Central Film Library approximately twenty-five programs of film society interest. After that it had to rely on other sources.

A second source was the commercial distributor. Some had films, already censored but not widely distributed, available for rent to a

[65] *Journal of the Film Industry*, July, 1952.
[66] Interview, C. Das Gupta.

film society. Unexpected treasures were found, such as the French *Un Carnet de Bal,* rented from an American distributor under the title *Life Dances On.* However, in the early 1950s the period of censorship licenses was changed from ten years to five years.[67] Hundreds of old films became automatically uncertified—and had to be recensored, with payment of fees, even to be previewed by a film society executive. The change led many commercial distributors to destroy prints of old films to save storage costs. Thus it had the effect of curtailing this source of supply.

A third source consisted of foreign embassies. As other sources dried, this became all the more important, especially because an embassy sometimes invited the society to its premises to see a film, averting costs and red tape of censorship. Film societies became especially dependent on embassies representing countries with nationalized film industries. The USSR could make available such films as *Alexander Nevsky* and *Ivan the Terrible,* whereas the U.S. Information Service could offer no feature films but only documentaries on such topics as the New York Public Library or Concord, Massachusetts.

From these various sources the Calcutta Film Society, during the years 1947–52, showed films from the United States, the United Kingdom, the USSR, France, Germany, Italy, Poland, Hungary, Czechoslovakia, Canada, Japan, and India itself. And it provided other valuable experience.

An early member of the Calcutta Film Society was Hari Das Gupta, a student at the University of Southern California in 1946–47. There he had met the French director Jean Renoir, who had come as a visiting lecturer. When Renoir later came to Calcutta to make *The River,* he turned to Hari Das Gupta for help, and eventually made him his assistant. Renoir spent months selecting locations. Sometimes Das Gupta, on these location-hunting jaunts, got Satyajit Ray to join them. Ray would take a long lunch period from the advertising agency, where he had become art director. He could talk endlessly—and beautifully—about Bengal and its people, and Renoir listened. Ray and Das Gupta learned from Renoir too. They were sometimes surprised at the familiar things that interested

[67] *BMPA Journal,* February, 1953. The ten-year period was restored in the 1960s.

Renoir and roused his enthusiasm and loving attention. Hari Das Gupta recalls, "We started seeing through his eyes."[68]

Renoir visited and talked to the members of the Calcutta Film Society. And there were others. When Pudovkin and Cherkasov visited India in 1951, they too met with the film devotees of the society. Later came John Huston and others.

From 1952 until 1956 the Calcutta Film Society suspended its activities. During that time its members talked unceasingly about reviving it, but the problems seemed formidable. In 1955 they learned, from the magazine *Indian Documentary,* that Bombay was having similar troubles:

> The Bombay Film Society has been unable to present its regular monthly showing to members for the past two months, due to obstacles and delays put before it by the Home Department of the Bombay government. Difficulties of exhibition and censorship are one thing; equally discouraging is the difficulty of importing into India, in the first instance, films destined for noncommercial exhibition to small groups scattered throughout the country. These various problems require study and sorting out by some central body of determined and enthusiastic people, before any real progress in the film society movement can be expected.[69]

The Bombay society did not meet again for two years. Thus by 1955 both pioneer societies had come to a halt.

The Calcutta Film Society had meanwhile played a vital role in the lives of several people. Satyajit Ray was in production. So was Hari Das Gupta. They would be followed soon by Chidananda Das Gupta and others. But a chronicle of the troubles of the film societies raises a question. What did independent India gain by the obstacles it put in their path? Did India not gain more from the desperate efforts of the societies to continue, in spite of massive official discouragements?

[68] Interview, H. Das Gupta.
[69] *Indian Documentary, July–September, 1955.*

Feud

We have briefly portrayed the three major film centers—Bombay, Madras, Calcutta—during the first postwar decade. Each had its own problems and was developing its own characteristics. But in some respects they were alike. During these years they shared a continuing feud with the government. Several aspects of this struggle need to be examined. One was the dispute over "approved" films. On both sides, this generated heat.

When independent India reintroduced the compulsory showing of "approved" films, various industry spokesmen denounced the action on principle. They said the central government had been "ill-advised by those trained under British bureaucracy,"[1] and argued that exhibitors would be willing to show documentary films and newsreels voluntarily. The record hardly supported them. After the termination of the *Indian News Parade* in 1946, a private Indian company had purchased its assets and tried to carry on the service. The effort had survived only a few months because exhibitors were uninterested.[2]

The industry spokesmen had apparently forgotten, or were unaware, that as early as 1937 the Motion Picture Society of India had itself urged the government to require exhibitors to show a minimum quota of educational film in each program, in the interest of "national culture."[3] At that time producers clearly felt that some exercise of authority would be needed to create a market for what was considered essential, but had become unmarketable.

While continuing to object to compulsory showings,[4] industry

[1] *Journal of the Film Industry,* February, 1950.
[2] *Report of the Film Enquiry Committee,* p. 51.
[3] *Journal of the Motion Picture Society of India,* January, 1937.
[4] There was some litigation on the subject. In Madras State the requirement, as incorporated in theatre licenses in 1948, made it obligatory to show "not less than 2,000 feet of one or more approved films." In 1954 the Madras State Court, in *Seshadri* v. *District Magistrate, Tanjore,* voided this on the ground that gov-

spokesmen also protested compulsory payments, as well as the scale of the payments. Exhibitors even argued that since the government paid newspapers for advertising space, it should pay theatres for documentary screentime, rather than require payments from them. The argument did not betray a high esteem for documentaries.

In 1951 the Film Enquiry Committee briefly considered the issue, and approved both the compulsion and the required payments. Meanwhile the Ministry of Information and Broadcasting took up the issue at various times in more detail.

Dr. B. V. Keskar, who became Minister of Information and Broadcasting in 1952, argued with some cogency that the required payments were a safeguard to the industry itself—necessary to protect the market for private producers. In December, 1954, he enlarged on this point in the Lok Sabha, the House of the People:

> The ultimate aim . . . is the development of the production of documentaries and newsreels by the Indian film industry. It has been acknowledged all over the world that documentaries, newsreels, and educational films must supplement the normal fare of entertainment given in the cinema houses and all the advanced countries have an established section of the film industry which produces films of this kind. . . . The only effective method of developing a documentary section in the film industry in India is to assure them a steady market, as no producer would otherwise be prepared to undertake the production of this kind of films which are essential to the community.[5]

Required showings with adequate payment, the minister said, were needed to create a market for documentary producers. If government began donating films of this sort to theatres, theatres would never again be willing to pay for them. The system of payments therefore protected the industry.

That its policies were designed to foster documentary film production in the private sector was a favorite theme of the ministry throughout these years. No other principle was so frequently reit-

---

ernment, in the absence of a stated maximum, would be able to preempt all theatre screentime. The state then rewrote the clause to require the showing of not *more* than 2,000 feet of approved film; other states followed this example. The requirement has remained in effect in this form.

[5] *Lok Sabha Debates,* December 9, 1954.

erated. And yet there is no doubt that some of the actions of the ministry were at the same time having an exactly opposite effect.

When the obligation to show approved films had been reintroduced and the Films Division of the Ministry of Information and Broadcasting established, it sent representatives to cinemas throughout the country with a contract—ready for the signature of the exhibitor. It was a block-booking and blind-booking contract, under which the Films Division undertook to provide all the approved films the theatre would need to fulfill its legal obligations for the year, and the theatre committed itself to show them and to pay for them. Every cinema in the country felt it had no choice but to sign the "contracts which are described as agreements," but which were "thrust by one party on the other."[6]

It was certainly true that the government had created a market for documentary films and newsreels by making the showing of such material mandatory and requiring payments for it. It was also true that the government had, a moment later, completely preempted that market for its own products. Ministry spokesmen, in defense of their action, said that theatres were free to purchase additional documentaries from the private sector, but their faith in such voluntary action was hardly more convincing than the industry's own.

The ministry, in implementing the law, had in effect gone far beyond the law. In theory a theatre had to show an approved film either from the private sector or the public sector. The government had in 1949 set up a Film Advisory Board to review films and approve those that served a public purpose, whether produced by the Films Division or by others. In actuality the ministry, through its block-booking contract, had rewritten the obligation imposed on theatres. In every program, every theatre in India was now obligated by contract not merely to show an approved film but to show *an approved film from the Films Division.*

The gist of the situation was that a privately produced documentary, though approved by the Film Advisory Board, was unmarketable unless the Films Division itself chose to buy it, at its price, for its distribution. An independent documentary producer might

[6] *BMPA Journal,* June, 1949.

conceivably survive as a supplier to the Films Division, but he was prevented from being its competitor. Because of the contract, a theatre would never be in a position to make a choice between an approved film from the public sector and an approved film from the private sector.

In answer to charges of monopolistic practices, Dr. Keskar, the Minister of Information and Broadcasting, offered in 1954 to negotiate a formula by which the preempted screentime would be divided with the private sector.[7] S. S. Vasan, as president of the Film Federation of India, rejected the idea of any such negotiations, so it is not clear what sort of arrangement the minister had in mind. The block-booking contract continued undisturbed.

The elimination or prevention of competition through block booking is a practice which India had known something about for decades. India was aware of the role it had played in consolidating the international position of the American producers after the First World War. It was aware of the role it had played in furthering their control over the American market. It is ironic that in the very year when block booking was outlawed in the United States by its Supreme Court, as the climax of an antimonopoly prosecution, independent India moved to create a closed production-distribution system through the same technique, but in a more thorough manner. Not a cinema in India remained outside the controlled market.

It should be emphasized that this controlled market in theatrical documentaries and newsreels was brought about *not* by the approved film plan, which envisaged competition, but by a contract for which there was no provision in the law. It was administration, not legislation, that created the monopoly.

Throughout this period the Ministry of Information and Broadcasting insisted that it was fostering the independent production of documentaries, and it apparently believed it was doing so. Not only did its Films Division occasionally buy a completed film on a lump-sum basis. It also had a policy of commissioning a few films each year to outside producers. Up to the end of 1954 six films had been commissioned in this way.[8] The practice would be expanded, the

[7] *Ibid.*, November, 1954.
[8] *Lok Sabha Debates,* December 9, 1954.

ministry said. Each year the Films Division issued a list of "approved" independent producers—at first about two dozen, later many more—who would be invited to submit competitive bids on film topics designated for outside production.

But the rates which the Films Division was willing to pay for the outside productions were a constant issue. This had erupted into public debate as early as 1950. By that year, according to industry estimates, the Films Division itself was spending Rs. 27 per foot in the production of staff-produced films. Yet it considered Rs. 12 per foot a proper price for private producers, except under extraordinary circumstances.[9] The negotiations on this subject were described as follows by the *BMPA Journal:*

> The Ministry was originally prepared to pay only Rs. 8 to Rs. 15 per foot according to the quality of the documentary and even then for all the rights involved, viz., theatrical, nontheatrical, classroom, television, etc., whether in India or abroad. After prolonged discussions the Ministry offered a fixed payment of Rs. 12,000 per documentary, and the documentary producers accepted it as a minimum payable against delivery of the negative of a film, plus the residue from the exploitation of the film in the world's 35 mm and 16 mm markets, because at Rs. 12,000 the producers would make a loss which could be covered only in the manner suggested. This compares with . . . the average of Rs. 27,000 spent by the Films Division itself per reel. The Films Division does not accept the figure quoted for it, but it does not reveal its actual expenditure. . . . Up to date the Ministry has not accepted the offer of the private producers.[10]

It was apparently felt by the government that a governmental production agency needed to spend far more than a private producer to produce comparable material. It had, it frequently insisted, administrative costs. The negotiations collapsed and prices continued to be set mainly by competitive bidding. Exceptions were made only for an occasional "prestige" film; for *Gotama the Buddha,* marking the twenty-five-hundredth anniversary of the birth of the Buddha, a contract was made with Bimal Roy without competitive bids, at approximately twice the usual price-per-foot for outside productions. In most cases the independent producers were—in the words used by several—"encouraged to cut each oth-

[9] *BMPA Journal,* August, 1950. [10] *Ibid.*

er's throats." Each year new names were added to the list of approved producers who were allowed to make bids, and other names dropped. Strenuous efforts among producers to organize a gentleman's agreement to put a floor under the bidding were frustrated by the eagerness of newcomers to break into production. Some producers spent several years on the list without winning a contract, but no doubt contributed to maintaining the low level of contract rates.

Fortunately for independent producers with documentary interests, there were occasional sponsors other than the government. In the early 1950s Burmah-Shell decided to launch a film department and engaged James Beveridge, previously of the National Film Board of Canada, to organize it. Under Beveridge, Burmah-Shell announced a policy of commissioning its films to Indian producers. Because of the notable work done under Shell auspices in other countries, the project aroused high hopes, and proceeded to justify them. Over a period of several years Burmah-Shell issued an impressive series of films including *Textiles,* directed by Paul Zils and winner of a 1956 Edinburgh film festival award; *Village in Travancore,* directed by Fali Billimoria and a winner at Edinburgh the following year; and *Panchtupi: A Village in West Bengal,* made by Hari Das Gupta—who had been assistant to Renoir on *The River.*

After this encouraging and even inspiring start, and amid indications of long-range planning, the Burmah-Shell film activity abruptly halted. The department was dissolved; James Beveridge left for home. No specific reason was given by Burmah-Shell, but various explanations were circulated. One was that Burmah-Shell had hoped for theatrical showing of some of its documentaries as approved films, but that Film Advisory Board policy had excluded consideration of any material carrying a commercial credit line, and that virtually no other distribution channels had been found available in India. Another explanation was that Burmah-Shell had been required by "government pressure"[11] to cut prices and public-relations expenditures. Another was that raw-film shortages were beginning to hamper the unit. In any event, the most encouraging

[11] *Indian Documentary,* Festival Number, 1958.

patron of the independent producer suddenly vanished from the film scene. At about the same time the United States Technical Co-operation Mission began to curtail its film production program, which had employed independent producers for several years. Documentary producers suddenly found that virtually the only source of employment was the Films Division, and that employment continued to be on a subsistence level, on competitive-bidding terms. In 1958 the *Statesman* aptly wondered whether the so-called independent producers should not properly be called "dependent producers."[12]

That year brought an atmosphere of crisis among these documentary producers. The magazine *Indian Documentary,* which had been their rallying center, titled an editorial: "IS IT FAREWELL TO DOCUMENTARIES?"[13] With the following issue it suspended publication.

It was not the end, however. There was some increase in the use of advertising films and slides in theatres, and some of the producers survived on these, between Films Division assignments. Occasional sponsors of public-relations films also continued to turn up. Hari Das Gupta and Fali Billimoria became outstanding producers in this special subdivision of documentary. A few states also sponsored occasional films. But these other sources of support were irregular or barren of incentive. For most documentary producers, best hope lay in having a place on the approved list of the one all-important patron, the Films Division, and doing its bidding, at its price, in its manner. In India the future of documentary film, as of newsreel, hung on the Films Division. The film of fact and the Films Division had become virtually synonymous.[14]

Meanwhile, during these years of conflict, the Films Division had grown from humble beginnings to an organization of size and sub-

[12] *Statesman,* March 9, 1958.

[13] *Indian Documentary,* Festival Number, 1958.

[14] The Films Division's own publications seem to assume this. The brochure *Films Division* opens with these words: "The factual film plays a significant role in India today. It is the most effective medium for dissemination of information and education to the masses. Established in 1948, soon after Independence, Films Division is now one of the largest short film producing organisations in the world." *Films Division,* p. 3.

stance. It included in its ranks a number of interesting and talented people.

Here and there in the Indian film world are individuals who bridge decades of film history. Such a person is Ezra Mir. Born in Calcutta in 1902, he attended St. Xavier's College, had a brief fling at business with his father, then managed to get a foothold as an actor in Madan Theatres Ltd. In that same year, 1923, he held a winning ticket in the St. Leger Sweepstakes and found himself in possession of Rs. 8,000. He bought some new clothes and booked passage for the United States. Arriving in New York, he got a job as an extra with Rudolph Valentino, who was making *The Sainted Devil* in a studio on Long Island. This helped Ezra Mir get the rest of the way to Hollywood, his real destination. After a year and a half of unemployment, with only intermittent earnings at odd jobs, he won a place as assistant cutter at Universal Pictures Corporation, and found himself near the bottom of a vast hierarchy. Hearing that Mr. Laemmle, the man at the top of this hierarchy, was contemplating purchase of the Madan chain in India, he got word to Mr. Laemmle that a young Indian was available who had information on the Madan organization. Promptly Ezra Mir was spirited to the Laemmle sanctum. It turned out that Mir could not answer any of Mr. Laemmle's questions about Madan's capital investments, but Mir took the occasion to let Mr. Laemmle know, with overwhelming earnestness, that he wanted to be a writer, not a cutter. Almost casually, he was moved to the script department. Later Mir worked for other Hollywood companies and wrote treatments for Dolores Del Rio. Then the coming of sound brought upheaval. Mir stayed on for a while, then felt it would be wise to return to India. He arrived in time for the great Indian transformation to sound. In Calcutta he found Madan in rapid decline so he moved to Bombay. There he at once became The Man Who Had Been in Hollywood When It All Started. He became a director of sound films for Sagar and later Ranjit and others. When war came, his international background made him a logical candidate for Information Films of India. He became its chief producer. Years later, as successor to M. D. Bhavnani and J. Bhownagary, he stepped into a similar position with the Films Division of independent India.

With the title producer-in-charge, he became the artistic supervisor of a complex, rapidly growing apparatus.

Few organizations have faced a more staggering production task. With headquarters in Bombay, it began by making about 36 documentaries per year for theatres—which soon grew to 52 per year and then 104 per year. From the start it maintained a pace of approximately 52 newsreels per year, which it soon supplemented with monthly "news magazines." It began by making each film in five languages—Hindi, Bengali, Tamil, Telugu, English—but later added Kannada, Malayalam, Marathi, Gujarati, Punjabi, Oriya, Assamese, Kashmiri.[15] Print orders, which stood at 68 prints for each film in 1951, gradually rose into the hundreds. It established a newsreel cameraman in each state, with additional men in the large cities. With assistance from the U. S. International Cooperation Administration it started a cartoon division. In addition to "approved films" for theatres it made special films for other purposes, including films for rural showings through mobile vans on such topics as health and planned parenthood; films for school use; and films for use at Indian missions abroad.

A meeting in New Delhi each year began the task of shaping an agenda for later months. At this meeting representatives of various ministries would meet with the Films Division to propose topics, review topics proposed by others, and classify them according to priority. Since the birth of the Films Division the overwhelming majority of topics has originated with ministries of the central government. A topic has usually had a ministry as its sponsor and has had a content consultant designated by the ministry. Even topics proposed by the Films Division itself, or by others, have been handled in this manner. Under the producer-in-charge and his deputy producer, the Films Division developed an echelon of assistant producers, each supervising several directors. A film ultimately became the task of a director—or, in some cases, of an outside producer. In most cases the staff director or outside producer has done the writing. At various stages he would have to check with the assistant

[15] In 1960 Urdu, a Hindi variant, replaced Kashmiri as state language of Kashmir.

producer, who in turn might check with the deputy producer or producer-in-charge, who might check with the ministry consultant. At several prescribed stages—script, edited workprint, test print—there would be meetings of director, assistant producer, deputy producer, producer-in-charge, and ministry consultant, to review, negotiate, correct, approve, or reject. A four-month production cycle became normal.

That an apparatus of such size, and with diplomatic proceedings so exacting, should function at all and meet its schedules is in itself remarkable. That its procedures should impose on the films a characteristic Films Division pattern and style is not surprising. From the very beginnings of the system, the films were under the control of ministry representatives with little or no film background. Some were men of considerable education, products of a highly verbalized culture. To them it was, quite naturally, the words in the narration that counted. The pictures—subsidiary, in their view—that would accompany those words could safely be left to others. The typical Films Division film has had constant narration, crowded with information. If the facts were there, embedded in the words, the consultant would usually feel his mission had been achieved.

The problem of issuing each film in numerous language versions contributed to the overemphasis on narration. Inevitably, dialogue was discouraged. Speeches by cabinet ministers, even those by Nehru, were often filmed without sound, and the gist of things said was given via narration, in the language of the particular version. Even if a minister's voice was used, it would be only a momentary use, with the voice quickly faded down for the narrated translation. Inadvertently, these procedures may have bolstered the national status of government leaders, in that a too-strong identification with any region was avoided.

The Films Division, in its distribution, tried to follow the policy of matching the language of the documentary or newsreel to the language of the scheduled feature. A Telugu feature would generally be joined by a Telugu documentary or newsreel. With features in circulation in numerous languages, the logistics problems involved in such a policy have been considerable.

Radha and Krishna, 1957. FILMS DIVISION PRODUCTION USING HISTORIC ART

This Our India, 1961. FILMS DIVISION CARTOON FILM

Pilot Project, 1962. FILMS DIVISION DOCUMENTARY ON MODERNIZATION OF RIVER TRANSPORT

Ironically, English became and remained the operating language of the Films Division. After approval in its English form, a narration would be translated into the official Indian languages. In some of the Indian versions, it was found that the narration might run longer than in English. An informal rule therefore developed: the English narration must not use more than 80 percent of the available time. It has seldom used less.

With all these operational handicaps, the Films Division yet developed during the 1950s into an organization commanding respect. Its films were modest and factual in manner and well photographed. In international festivals they began to win "certificates of merit" and occasional other awards. At the 1951 documentary film festival at Venice, *Jaipur* (Jaipur) won a First Prize. At the 1956 festival of documentary and experimental films in Montevideo, *Symphony of Life* won a First Prize. At the Manila festival in the same year, *Khajuraho* (Khajuraho) won a Silver Carabao for "cultural values." The following year at Cannes, *Gotama the Buddha* was cited for "exceptional moral and artistic beauty."[16]

To be sure, the films continued to be a target for frequent criticism at home. One complaint was that too large a proportion of newsreel items involved activities of cabinet members—cutting ribbons, laying cornerstones. No doubt this was partly a result of the readiness of ministers to cut ribbons and lay cornerstones. Another charge was that this predilection for the activities of cabinet members favored the party in power. Yet it was hardly up to the Films Division to avoid cabinet members in deference to defeated parties. The films were also criticized for stodginess and sameness of manner. The biographical films were criticized for a tendency to sweep controversial issues under the rug.[17]

Yet with all these problems and criticisms, the Films Division was accomplishing something. The Film Enquiry Committee, in 1951, did not find audiences as hostile as exhibitors had claimed. The documentaries and newsreels were opening the eyes of filmgoers to many phases of the development of their country, and to the huge tasks facing it. The films on Indian sculptures and cave paintings were acquainting them with an artistic heritage which, to astonishing numbers of Indians, had long been a closed book. The geographical films were showing them parts of their own country which they had never seen and which, a few years before, they had not known existed.

The Films Division represented the first step of the government of independent India in "public-sector" film enterprise. Under the Indian five-year plans, the ultimate aim is a socialist economy. Certain basic fields have been allocated entirely and immediately to the public sector. In others, "private-sector" enterprise has been encouraged to continue, although "public-sector" enterprise may also be undertaken. Still others have been left largely to private initiative.

[16] *Films Division,* pp. 11–12.

[17] In a biographical film on the nationalist leader Lokmanya Tilak, for example, his disagreements with Gandhi were completely ignored, and the viewer left with an impression of complete harmony in the independence movement—an example of the hazards of official biographies.

The inauguration of the Films Division placed the film medium in the middle category. A few years later the government decided on a second step in public-sector film activity. The Film Enquiry Committee had found an almost complete absence of films serving the needs of children and had urged the film industry to consider those needs. The Ministry of Information and Broadcasting, in its 1954–55 *Report,* announced the decision of the government to "promote the formation of a Children's Film Society.[18] It was launched in 1955 as a quasi-independent corporation and was provided with government funds to produce and distribute children's films.

The government envisaged other possible steps in public-sector film enterprise. In 1955 Prime Minister Nehru addressed a seminar of leading film artists, held in New Delhi under the auspices of the Sangeet Natak Akadami (Academy of Music, Dance, and Drama). It was the first time the government had convoked a group of film makers in a "cultural" context. The seminar was conducted under the guidance of Devika Rani, who emerged from private life for the unusual occasion.

The Prime Minister described the film medium as a "tremendous thing," with an influence in India "greater than the influence of newspapers and books all combined" (CHEERS), but quipped: "I am not at the moment talking about the quality of the influence" (LAUGHTER). He then told the assembled film makers, all from the private sector, that the government would be "likely to compete with private ventures in films." He felt that "the result might be a setting up of standards by a certain measure of competition."[19]

This comment raises the question of whether "a certain measure of competition" might not also have been beneficial to the Films Division, an organization serving a controlled documentary market. The Prime Minister may not have been aware of how firmly his government had moved to avert the development of private competition in this field.

The central instrument of those moves had been the block-booking contract imposed on theatres soon after independence. Seldom has a contract conferred firmer control. Illustrative of this is the

[18] *Report, 1954–55,* p. 3. [19] *Film Seminar Report,* pp. 13–15.

curious arbitration clause of the contract. In the event of any dispute between the government—the "distributor"—and the party of the second part—the "exhibitor"—the contract provided that the dispute would be "referred to the sole arbitration of the Secretary, Ministry of Information and Broadcasting, Government of India," or anyone designated by him. Apparently aware that exhibitors might consider this a one-sided arrangement, the contract added: "No objection shall be taken to any such reference on the ground that the person so appointed is a servant under the Distributor and has to deal with the matter in the course of normal duties."[20] In other words, all disputes would be settled by the government.

By the end of its first decade the Films Division had grown into an organization of over five hundred employees, with its own compound, its own group of buildings, its own nationwide distributing organization. It had become one of the few elements of stability in the Indian film world. Most of its technicians, while low-paid, had a slightly higher level of pay than comparable workers in the private sector,[21] and had in addition a measure of security. Under the leadership of the patient, ever-diplomatic Ezra Mir, producer-in-charge, the Films Division was winning the strong confidence and support of government. An air of assurance was beginning to surround the organization.

In 1957 the Ministry of Information and Broadcasting decided on a new move. In the chief cities it would inaugurate special theatres showing exclusively government documentaries and newsreels, for a very low admission charge. The first such venture was the Films Division Auditorium in the heart of New Delhi, opened in 1957, with an admission price of Re. 0.25. It was the ministry's first cautious move to test the independent box-office attraction of its productions, so long deprecated by exhibitors. The small auditorium was opened with appropriate ceremony by R. K. Ramadhyani, Secretary of the Ministry of Information and Broadcasting—later presidential secretary. Ramadhyani stated: "In this auditorium it is

[20] *Agreement,* p. 3. Contracts between the Films Division and independent producers had a similar "arbitration clause."
[21] *Report of an Enquiry into the Conditions of Labour,* p. 49.

proposed to have two shows of films, one in the forenoon and one in the afternoon, of about one hour each. Depending on the demand, the number of shows may be suitably increased."[22] The resulting attendance almost immediately caused the number of shows to be "suitably" decreased. The forenoon showings were dropped entirely. Monday showings were eliminated. Sunday shows were lengthened to two hours. In a city of a million, capital of the nation, the official documentaries could hardly draw a thousand people in a week.

As early as 1950 the Ministry of Information and Broadcasting was estimating that its documentaries were being seen by 15 million people.[23] Its auditorium venture, though described in an official report as "very popular,"[24] must have shown the ministry how dependent it was for its mass audience on an industry whose products it heartily disliked. If each of the documentaries was reaching 15,000,000 people, it was doing so as a hitchhiker, riding with feature films. But this strong dependence of the public sector on the private sector was not likely to increase the affection between them—as we shall see.

There are other kinds of music

When Dr. Keskar—Balkrishna Vishwanath Keskar, D. Litt.—became Minister of Information and Broadcasting in 1952, he acquired responsibility for a number of government agencies dealing with the mass media. They included among others the Films Division, the Central Board of Film Censors, the Mobile Units, the Publications Division, the Press Information Bureau, and the entire broadcasting chain known as A.I.R.—All India Radio.

Dr. Keskar had not had experience in any of these media, but he brought to his task an impressive record. Educated at Poona, Hy-

[22] Ramadhyani, *Essays and Addresses,* p. 206.
[23] Statement by R. R. Diwakar, quoted in *Journal of the Film Industry,* January, 1950.
[24] *Report, 1959–60,* p. 20.

derabad, Banaras, and Paris, he had long been prominent in the councils of the Indian National Congress. In 1946 he had been a member of India's Constituent Assembly. In the same year he was appointed Deputy Minister of External Affairs. In 1950 he served on the Indian delegation to the United Nations. Two years later he became Minister of Information and Broadcasting.

He had always been a devotee of Indian classical music, and this gave him an immediate special interest in All India Radio, which he envisaged playing a prominent role in a revival of Indian classical culture—especially its music.

In relation to the film industry Dr. Keskar seemed less at ease, but he soon began to make official appearances at film functions. On these occasions he generally wore the *chudidar,* the tight pants worn by many Congress leaders, along with the *sharwani* or long coat. In this native garb, his thin figure had an ascetic look in almost any film gathering.

Film makers soon learned that Dr. Keskar had many misgivings about film and the film industry and would be frank about them, whenever asked to speak. Soon after taking office he told a conference of producers: "Producers in this line should have a certain background of culture. . . . At present there is hardly any standard maintained by many of the productions we see on the screen."[25] A few months later, at a tea party given for him by the Bengal Motion Picture Association, he told the gathering: "My experience of the film industry has been that a large part of it is not conscious of its proper functions, although I know that there must be many at the same time who are."[26]

Meanwhile, on several occasions, Dr. Keskar had criticized film songs, and had already made policy decisions that reflected his feelings about them. In July, 1952, the *Journal of the Film Industry* carried the news that All India Radio would reduce the time given to film songs.[27] Dr. Keskar had also decided to end the long-standing practice of mentioning the titles of films from which the songs came.

[25] *BMPA Journal,* July–August, 1952.
[26] *Ibid.,* May, 1953.
[27] *Journal of the Film Industry,* July, 1952.

He would permit the name of the singer, but the name of the film would be considered "advertising," and disallowed. Dr. Keskar did not heed the angry protests of the producers—who were the copyright owners—with the result that they decided to discontinue the performing licenses under which All India Radio had been broadcasting film songs. Thus film music abruptly disappeared from all transmitters in India.

This sudden purification of the airwaves was, to many devotees of Indian classical music, a major event and a triumph, for which Dr. Keskar was warmly praised. For they regarded film music not only as an abomination but as a threat to a sacred cause—the survival of classical music.

Their feelings on this subject touched on many issues, but began with that of instrumentation. The use in film songs of such instruments as the piano, harmonium, vibraphone, xylophone, and saxophone involved adoption of the Western tempered scale. This, it was felt, was rapidly blunting Indian ears to the nuances of traditional Indian music. During the 1955 Film Seminar—also addressed by Nehru—this point was discussed by R. Ranjan, a prominent actor, dancer, and musician, and director of a school of music and dancing in Madras:

> With the adoption of the tempered scale of the west, our musicians become oblivious of the delicacies and subtleties of the 22-sruti scale. . . . What gives eternal strength and charm to Indian music is its immense potentiality for delicate touches by the use of the 22-sruti scale.[28]

For this reason, he felt, the survival of Indian music required drastic action. Moves against film music were only a part of what was needed. He called for official steps to "popularize correct ideas" about musical scales and instruments and to "ban the use of the harmonium and other keyboard instruments from all schools, col-

[28] *Film Seminar Report,* p. 155. The use of quarter tones and microtones is a distinctive feature of Indian classical music. Twenty-two such notes or *srutis* (ten notes in addition to the universal twelve semitones) have been the foundation of the Indian musical scale. In brief, the octave is divided into twenty-two unequal intervals.

leges and dance institutions."[29] He was willing that the violin should stay; it placed no limit on nuances and, moreover, its origin had been traced to India. But the rest "must vanish immediately."[30] Then there could be "a concord of sweet sound . . . which would entrance our inner spirit."[31]

Such feelings were all part of the larger crusade—so important to Dr. Keskar—on behalf of Indian classical music. Under him more than 50 percent of all broadcast music came to be Indian classical music. Indian folk music, played in traditional styles, became another large classification. All India Radio accumulated a list of "over 7,000 approved classical music artists,"[32] who were to make intermittent appearances, and each year the list was increased. In a talk broadcast in January, 1957, Dr. Keskar reviewed the efforts made:

> The object is to encourage the revival of our traditional music, classical and folk. Both were in a state of decay and somnolence. It is obvious that music, which formerly flourished on account of royal and princely patronage, will not revive and flourish unless the State can extend to them the same or extended patronage. The Radio is fulfilling that task for the nation and I can say with satisfaction that it has become the greatest patron of Indian music and musicians, greater than all the princely and munificent patronage of former days.[33]

All India Radio was acclaimed by many as the savior of Indian classical music. It was largely in the regime of Dr. Keskar that it won this name.

One of the functions of the Ministry of Information and Broadcasting, of which All India Radio is a part, has been to serve as a "link between the Government and the people." One of its tasks, in the era of newly won freedom, was to "rouse in the common man a new sense of urgency and duty to the community."[34] In the film medium, the ministry had the Films Division for this task; in the radio medium, All India Radio. But in the latter case the entire system, including all transmitters and programing, was controlled

[29] *Film Seminar Report,* p. 155.
[30] *Ibid.,* p. 228. [31] *Ibid.,* p. 155.
[32] *Report, 1956–57,* p. 69. [33] *Ibid.* [34] *Report, 1954–55,* p. 1.

by the ministry. For radio in India had been nationalized under the British as early as 1930 and had been built up in the image of the British Broadcasting Corporation. With the coming of independence, it naturally remained in the "public sector."

At that time—August, 1947—All India Radio had six stations and less than a quarter of a million licensed listeners.[35] The small size of the following was accepted as natural in view of the close identification the system had had with British rule. Its prewar Director-General, Lionel Fielden of the British Broadcasting Corporation, had himself described his four years of strenuous effort as "enough to make a cat laugh. It was the biggest flop of all time."[36]

But independent India foresaw a major role for radio in the era of national transformation, and embarked on an A.I.R. Development Plan designed to make the system "available to a population of about 220 millions"[37] out of India's four hundred million. The first decade of freedom saw a substantial expansion of the facilities. By January, 1957, the six stations had grown to 28 stations, offering primary coverage to most of the nation. All major language groups were being served. The expansion had been accompanied by government drives to promote the sale of radio sets. But with all these efforts the number of licensed sets had grown, by October, 1956, to only 1,128,599.[38] Only a small proportion of these were community receivers. Most of India's population was still untouched by radio. Dismaying as the statistics were, still more discouraging was the fact that most of the existing sets were tuned regularly—especially in the late afternoons and early evenings—to Radio Ceylon.

The exact dimensions of this problem were not made clear. The Ministry of Information and Broadcasting itself made an audience study during this period,[39] but declined to reveal its findings. Five years later inquirers were still told by the ministry that the study

[35] *Report, 1956–57*, p. 24.
[36] Fielden, *The Natural Bent*, p. 204.
[37] *Report, 1954–55*, p. 1.
[38] *Report, 1956–57*, p. 26.
[39] *Report, 1957–58*, p. 21.

was not "public information."[40] Meanwhile less formal checks by foreign broadcasters and commercial advertisers indicated that the commercial short-wave service of Radio Ceylon, with a schedule consisting solely of Indian film songs and advertising, dominated the air over India at peak hours.

Dr. Keskar tried to minimize the issues raised by film songs and the rise of Radio Ceylon. In October, 1954, he was quoted as declaring that "except for raw and immature people like children and adolescents,"[41] householders in general detested film music.

Light was later thrown on the situation by UNESCO, which gathered and tabulated statistics on radio sets in various countries "around 1958." These figures showed:[42]

	Radio sets per 1,000 inhabitants		*Radio sets per 1,000 inhabitants*
Japan	158	Laos	7
Ceylon	25	India	4
Indonesia	7		

When All India Radio had launched its drive for the "revival of our traditional music," there was no thought that this might conceivably limit the network's role as "link between the Government and the people." Yet indications were that this had happened.

Naushad Ali, a leading music director for films, responsible for a number of the most successful film songs, offered an explanation. Referring particularly to north Indian classical music, he wrote:

> Classical sangeet has never been the art of the masses. It was first born in the sacred temples and later flourished in the glamorous courts of the Rajas, Maharajas and the Nawabs. . . . The common people who had no access to the great durbars were never offered the opportunity of listening to classical music. They could not, therefore, acquire an appreciative ear for it.[43]

The attempt to make this highly specialized music a part of the everyday environment of millions was, to Naushad Ali, an artificial imposition. The music had been in the first instance the preoccupation of small elite groups, who took special pride in its mysteries. To many millions of Indians it was almost as remote as the music

[40] Interview, Baji. [41] *BMPA Journal,* October, 1954.
[42] *Developing Mass Media of Asia,* p. 59.
[43] *Indian Talkie, 1931–56,* p. 99.

of British string ensembles. In the view of Naushad Ali, film music—a spontaneous and exuberant growth, emerging from an older folk music and adapting itself to a new era and its influences—was the real folk music of modern India.

Throughout these years of debate, the uncompromising level of All India Radio's programming had a magnificence of its own. It introduced as an annual event a National Symposium of Poets—something few nations would attempt. It regularly broadcast readings in Sanskrit, a language spoken, according to government staitstics, by 555 people.[44] It organized a Music Symposium in which musicologists discussed such topics as "the Evolution of Dhruvapada in Hindustani Music" and "The Evolution of Kritis in Carnatic Music." The National Programme of Talks, broadcast over all stations, offered such subjects as "Modern Prose and Traditional Sanskrit Style," "Adequacy of Indian Prose for Contemporary Needs of Expression," and "Intellectual Life in Pre-British India." Current controversy was shunned; the 1956–57 *Report* mentioned with apparent satisfaction that "controversial party broadcasts have again been avoided." However, the Chief Election Commissioner delivered three national talks on "the desirability of maintaining law and order during the elections and the duties of public servants in connection with the elections."[45]

Despite its generally unflinching performance, it was not a great surprise when All India Radio in 1957 took a step in a different direction. This move had to do, in part, with film songs. The film producers had already renewed the performance licenses of All India Radio. This was now followed, with as much fanfare as the situation allowed, by the inauguration of a new service, "a landmark in the history of All India Radio." Over two powerful short-wave stations from Bombay and Madras, blanketing the nation, India began a continuous offering of "popular music and light entertainment." Portions of it were also to be rebroadcast, at various times of the day, over the regional medium-wave stations. The ministry made it clear that the new service would not offer *only* film music. There were, it was emphasized, other kinds of music. But film

[44] *India, 1961,* p. 23.
[45] *Report, 1956–57,* p. 17.

songs "approved by the Screening Committees" would be used, and "adequate allocation of time for film music has been made at each station."[46] The raw and immature, the adolescent and the child, were being invited back.

For the mass of the people

Throughout the struggles over film songs and documentary films, there was also conflict over censorship.

The determination to cleanse India of corrupting Western influences was a force in this conflict, as in the struggle over film music. The impulse showed itself in a variety of ways, including a determination to enforce stricter decorum in manners and dress. Although kisses, even between Indians, had occasionally been permitted in the 1930s—the royal lovers in *Karma* kissed several times—kissing became more strictly taboo after independence. For a while this taboo was enforced even in foreign films. Indians became accustomed to strange jumps in such films. A shot of lips approaching would be abruptly followed by a shot of lips withdrawing. Similarly, in a drinking sequence, a hand lifting a glass from a table would be abruptly followed by the hand replacing the glass on the table.

The absurdity of such effects led to occasional relaxation of standards for foreign films. This was rationalized on the ground that drinking and public kissing were customary among foreigners, and also on the ground that foreign films were now shown regularly in only a few dozen theatres, reaching only a relatively sophisticated segment of Indian society. This double censorship standard was always sharply criticized by Indian producers. The *BMPA Journal* complained bitterly: "Differential treatment to Indians and Englishmen had been the fate of India under British rule. Codes of treatment of one and the other in almost every sphere used to be different."[47] The *BMPA Journal* considered it scandalous that independent India should continue discriminatory practices. Ministry spokesmen were thus periodically pushed into promising equal treatment for all—which meant, in effect, equally harsh treatment for foreign as for Indian films. But strict adherence to this policy was found to be difficult.

[46] *Report, 1957–58,* p. 9. [47] *BMPA Journal,* May, 1950.

Although censorship in India was probably already more severe than that in any other leading film-producing nation, officialdom was under constant pressure to intensify rather than ease it. At virtually all sessions of the Lok Sabha and the Rajya Sabha, the two houses of the Indian parliament, members requested the Minister of Information and Broadcasting to pledge renewed effort to purge the film industry of unwholesome influences. A leader in these demands was Mrs. Lilavati Munshi, wife of the lawyer, author, and statesman K. M. Munshi. A member of the Rajya Sabha, she was also a constant letter writer and speaker in the cause of stricter censorship, and eventually formed a Society for the Prevention of Unhealthy Trends in Motion Pictures. She attacked the American and Indian film industries with equal indignation:

> There is hardly any Hollywood picture that does not show long and passionate kisses and hardly any Indian picture without a boy running after a girl. The dances are all so designed as to excite the lower instinct lurking in every human being. Newspapers and journals too give colourful stories about cinema stars to boost circulation. Some of them print pin-up girls to be viewed by impressionable adults. As a result, many young people leave their homes dreaming to become cinema stars. This disease is widespread even among very young boys.[48]

She addressed a letter to "Mrs. Mamie Eisenhower, White House, Washington, D.C." to tell her that some Hollywood films, not specifically identified, "come to India and ruin the moral fibre of our younger generations." If Mrs. Eisenhower could "stop the production of such Hollywood films," Mrs. Munshi told her, it would surely be "one of the monumental acts" of the Eisenhower presidency.[49]

State legislatures also agitated periodically over film morals. In 1958 the question of whether rock 'n' roll should be banned from Indian films was debated by the Madras legislature, without anyone seeming to know exactly what it was. After some debate the Home Minister, Mr. Bhaktavatsalam, was asked to enlighten the legislators on the nature of rock 'n' roll. He answered: "I do not know the

[48] *Report of the Society for the Prevention of Unhealthy Trends in Motion Pictures,* No. 1, p. 9.
[49] *Ibid.,* No. 2, pp. 12–13.

details or the technique of it, but I have heard it is an obscene dance performed by men and women."[50]

Indian producers were vaguely aware that film interests in the United States had reduced censorship interference by legal appeals based on constitutional guarantees. A few such efforts had been made in India—so far without result. The efforts prompted advice from Dr. Keskar: "I would warn the industry not to run after the mirage of getting asserted a particular right by legal means."[51]

The hope of successful anticensorship litigation did indeed appear to be a mirage. The legal framework within which censorship was operating in India differed sharply from that in effect in the United States. In the latter, governmental powers were circumscribed, soon after adoption of the federal Constitution, by guarantees of the Bill of Rights, including the First Amendment guarantee of freedom of the press. This guarantee, broadened by judicial interpretation to include film and broadcasting, has been a sort of Magna Charta of the media of expression. In the Republic of India the Constitution likewise acquired, soon after its adoption, a First Amendment dealing with censorship, but its effect was the opposite: it was a Magna Charta for the censor. The Constitution itself, in its Article 19, had established "the right to freedom of speech and expression." India's First Amendment, adopted in 1951, whittled this down by authorizing parliament to enact "reasonable restrictions" on the freedom of speech and expression "in the interests of the security of the State, friendly relations with foreign states, public order, decency or morality, or in relation to contempt of court, defamation or incitement to an offence." Thus censorship in India acquired an explicit constitutional base, which gave the government censorship powers difficult to challenge.

Official censors could act with assurance. In 1952 the Central Board of Film Censors listed transgressions that could result in censorship. Its list—later revised from time to time—included such colonial holdovers as taboos on "excessively passionate love scenes," "indelicate sexual situations," "unnecessary exhibition of feminine underclothing," "indecorous dancing," "realistic horrors of warfare," "exploitation of tragic incidents of war," "blackmail associ-

[50] *Mail,* September 4, 1958.
[51] Quoted in *BMPA Journal,* July–August, 1952.

ated with immorality," "intimate biological studies," "gross travesties of the administration of justice," as well as material likely to promote "disaffection or resistance to Government." Other items of a previous era, such as "scenes holding up the King's uniform to contempt and ridicule," had necessarily disappeared, although in a sense they were retained in broader form in a taboo on material likely to "wound the susceptibilities of any foreign nation." The new directives also borrowed from the Hollywood Production Code, as in the prefatory General Principles banning any film "which will lower the moral standards of those who see it."[52]

While the full range of authorized and proclaimed grounds for censorship appears to have been used by Indian censors, their actions seem, like literary trends, to have run in cycles. An obsession of the mid-1950s was described as follows by the *BMPA Journal:*

> Censorship in India is fast becoming a censorship of the female anatomy with the emphasis currently in vogue on cutting "Emphasized bosom" of heroines in some of our pictures. We deplore any attempt on the part of anyone to exploit the lower emotions of man but we cannot agree that the female anatomy should be tampered with to please the neo-moralist, that is the Indian film censor. We do not know whether there has been any new directive to the censor which is kept before the mind's eye while examining pictures for certification. The common boy or girl does not pay as much attention to the dress or contour of a woman as the censors do.

The editorial was referring to censorship orders of the following sort. For the Hindi film *Dara* (Dara):

> Delete. . . . Mid-close and close shots showing Usha with emphasized bust when she is on a jeep, as she jumps down, as she runs facing camera.

For the Telugu film *Pempudu Koduku* (Foster Child):

> Delete . . . the close-up of Sundai's busts when she is lying dead on the bed.

For the Tamil *Manitanum Mrigamum* (Man and Beast):

> Reduce close-ups and side-shots of Kamala's busts in the second dance.

For the Hindi *Gunehgar* (Sinner):

> Delete. . . . Close view of bust of Sarla as she is lying on back in the gangster's den.[53]

[52] *Indian Motion Picture Almanac and Who's Who, 1953,* p. 215.
[53] *BMPA Journal,* November, 1954.

A particularly arbitrary aspect of Indian censorship, after independence as before it, has been the sudden reversal—the abrupt uncertification of a film already approved and in distribution, and for which heavy promotion expenses may have been made. Thus in 1956 a number of feature films dealing with Africa, all of which seemed harmless to the censors when first reviewed, were suddenly uncertified. National susceptibilities, conveyed through informal diplomacy, appear to have been involved. The films included *The African Queen, The Snows of Kilimanjaro, Untamed, Tanganyika, African Adventure,* and *Below the Sahara.* The official explanation was that they "fail to portray the people of Africa in proper perspective."[54] A similar action was taken in regard to *The King and I.* Producers, distributors, and exhibitors felt they had no recourse against such reversals.

On the subject of censorship, as on other phases of the government-industry feud, protests and discussions accomplished little, and often only led to increased irritation on both sides. To the industry the censors were, by and large, bureaucrats intent on being as arbitrary as their British predecessors, and often more bigoted in their decisions. To many government people, the industry was an agglomeration of irresponsible, fly-by-night units, whose works seemed irrelevant to the great tasks of independence, who were intent on making money by the exploitation of sex and sensation, and who deserved to be dealt with firmly on moral questions—and taxed as severely as possible.

Dr. Keskar, in his appearances at film conventions, usually defended the need for strict censorship, and did so in terms which were especially irksome to film producers. He defended strict censorship as the will of the people, an expression of democracy. He appeared as champion of the popular will:

> Films in Indian languages are meant for and seen by the mass of the people, most of whom are not educated. . . . Now, the mass in any country is to some extent conventional, has certain prejudices that cannot be helped. An intellectual or educated audience can forgive or even appreciate unconventional themes or ideas put on the screen. The same cannot be said of the bulk of the people. I am afraid this fact is conveniently forgotten. . . .

[54] *Statesman,* May 5, 1956.

Unfortunately government cannot forget it because it is elected by the mass of the people and it has to take into consideration their feelings and sentiments.[54]

That Dr. Keskar, whose radio policies were facilitating the rise of Radio Ceylon, should be interpreting for film producers the "feelings and sentiments" of the masses seemed to producers particularly absurd and outrageous.

Approved films, film songs, censorship, taxes. In these and other areas of dispute—to some extent surface manifestations of a struggle between private and public enterprise—industry and government exchanged argument and challenge throughout the first decade of independence. To judge from trade publications, this feuding was the chief preoccupation of the world of film. Fortunately it was not.

Throughout the decade men were making films. Each year some three hundred films emerged on Indian screens. As had been the case each year since the Second World War, the three hundred films represented almost as many different producers. As had been the case in each of those years, the almost three hundred producers involved scores of newcomers—often called "adventurers." Among those who made their debut in the mid-1950s was Satyajit Ray, of Calcutta.

[54] Quoted in *BMPA Journal,* July–August, 1952.

Wide World

Like many of those called adventurers, Satyajit Ray began his first film with only a fraction of the funds needed to finish it. And like many, he began without any film experience, either as cameraman, director, producer, performer, or assistant of any sort. Yet there had been a preparation.

He was born in 1921 into an extraordinarily gifted family. His father, Sukumar Ray, was a prominent Bengali writer as well as a painter and "master of the photographic art." Sukumar's writings included children's stories and nonsense verses "which have come to stay as permanent stock in our juvenile literature."[1] These were printed in a children's magazine which he and his father had founded and which they printed on their own printing press and filled with joyful and imaginative drawings. The grandfather, Upendrakishore Ray, was not only a prolific writer and artist but also a violinist and a pioneer in half-tone block-printing in India. He was a friend of Rabindranath Tagore, who came frequently to the Ray home.

Sukumar carried on the children's magazine after Upendrakishore died, but was often in ill health. Satyajit would always retain an image of his father in an oxygen mask, and near death, still dictating a nonsense rhyme. Satyajit was not yet three when his father died.[2]

The death put the family in financial crisis. The printing press which they had operated for decades had to be abandoned. Satyajit grew up in the home of an uncle, while his mother taught embroidery and leatherwork in a home for widows. These trials may explain why Satyajit, on entering college, studied economics, in which at nineteen he earned a B.A. from the University of Calcutta.

[1] Sen, *History of Bengali Literature,* p. 341.
[2] For Ray's "ancestral tapestry" see especially Seton, *Portrait of a Director: Satyajit Ray.*

RABINDRANATH TAGORE.
COURTESY *The Hindu*

Rabindranath Tagore had long taken an interest in Satyajit, urging his mother to let him study at Santiniketan. As a result he went there in 1940 for further studies under Tagore, staying until 1942. Tagore, who had influenced almost all the arts of modern India, died in 1941.

Santiniketan represented resistance to the traditions of rote learning. Here the emphasis was on development from within. The plan included daily meditations and group meetings outdoors in a garden atmosphere. At Santiniketan Satyajit Ray concentrated on study of the graphic arts.

In 1943, at twenty-two, he entered the Calcutta branch of D. J. Keymer, a British-owned advertising agency, to earn his living as an advertising artist. Four years later he became art director of the branch.

Throughout his years of study he had been an ardent filmgoer. In his teens he had already selected films according to their directors rather than their stars. He wanted to see films directed by John Ford, Ernst Lubitsch, William Wyler. He also read every available book about film, including works of Eisenstein and Pudovkin. He studied the scripts in *Twenty Best Film Plays,* the 1943 anthology compiled by John Gassner and Dudley Nichols. On a few occasions he watched his cousin, Nitin Bose, direct at New Theatres. In 1947, the year of independence, Satyajit Ray and Chidananda Das Gupta founded the Calcutta Film Society largely as a vehicle for continued, intensive study of film.

Satyajit Ray already harbored thoughts of film production and for some time pursued an exacting method of training. When a film adaptation of a well-known work was about to appear, he would study the book and write a complete film script. Watching the produced film, he compared it inwardly with his own version, noting opportunities he might have missed and matters on which he would have improved on the produced film. By this technique he gained knowledge of his medium—and mounting confidence in himself.

In 1950 the Keymer advertising agency decided to send him to London for a period of study and training at the head office. The trip also proved an extraordinary opportunity to see films. In London he saw new and old films of many lands, not available in India, and became especially excited over the work of De Sica, Visconti, and other Italian directors.

When Ray returned to Calcutta, Jean Renoir—with whom he had become acquainted before the trip—was hard at work on *The River.* The chance to watch several days of location shooting formed the next step in Satyajit Ray's film education. With Renoir, Ray discussed a plan forming in his mind, and received strong encouragement to go ahead.

Throughout the years at the D. J. Keymer agency Ray, as a side occupation, illustrated books and designed book jackets. In the course of this work he designed a new, abridged edition of a widely read Bibhuti Banerji novel, *Pather Panchali* (Song of the Road). Not surprisingly, a screen version took shape in his mind, and this idea became an obsession. He sounded out Banerji on the possibility of turning the novel into a film.

A number of film producers wanted to buy the screen rights. When the author died, the problems of selecting a producer and negotiating a sale were left to the heirs. The illustrations made by Ray, and his obviously deep understanding and love of the work, helped the heirs to decide. Although Ray had never produced a film, he was permitted to buy the screen rights to *Pather Panchali* for Rs. 6,000.

As he worked on his screenplay, he also began to look for people and locations. In the tradition stemming from Robert Flaherty and more recently exemplified by the Italian neorealists, Ray was intent on using natural backgrounds as much as possible, as well as making maximum use of nonactors. He especially wanted to avoid familiar star faces, which would tend to shape roles into the molds of previous successes. Gradually the pieces began to fit together. He found a village, a meadow, a patch of woods, a boy, a girl. He tentatively selected a cameraman and technical assistants. The most elusive problem, for a long time, was the casting of the ancient aunt.

Pather Panchali became a feverish adventure that consumed all after-hours, weekends, and holidays. As production planning progressed, it was accompanied by the search for funds. Like many another, Ray began to seek out potential backers, particularly film distributors. Before long he had called on several dozen financiers. The encounters followed a pattern. The door was in each case open to him, for he held the screen rights for *Pather Panchali,* which many had wanted. They were interested in knowing his plans and proposals.

To explain these Satyajit Ray had a notebook in which he had written his entire screenplay, specifying camera usage and even including on each page a series of sketches, to indicate the composition of key shots. In an accompanying sketchbook, dramatic highlights had been pictured in greater detail with wash sketches.[3]

[3] The script and book of wash sketches have been donated by Satyajit Ray to the Cinémathèque Française, Paris. For examples of the wash sketches, see the title pages.

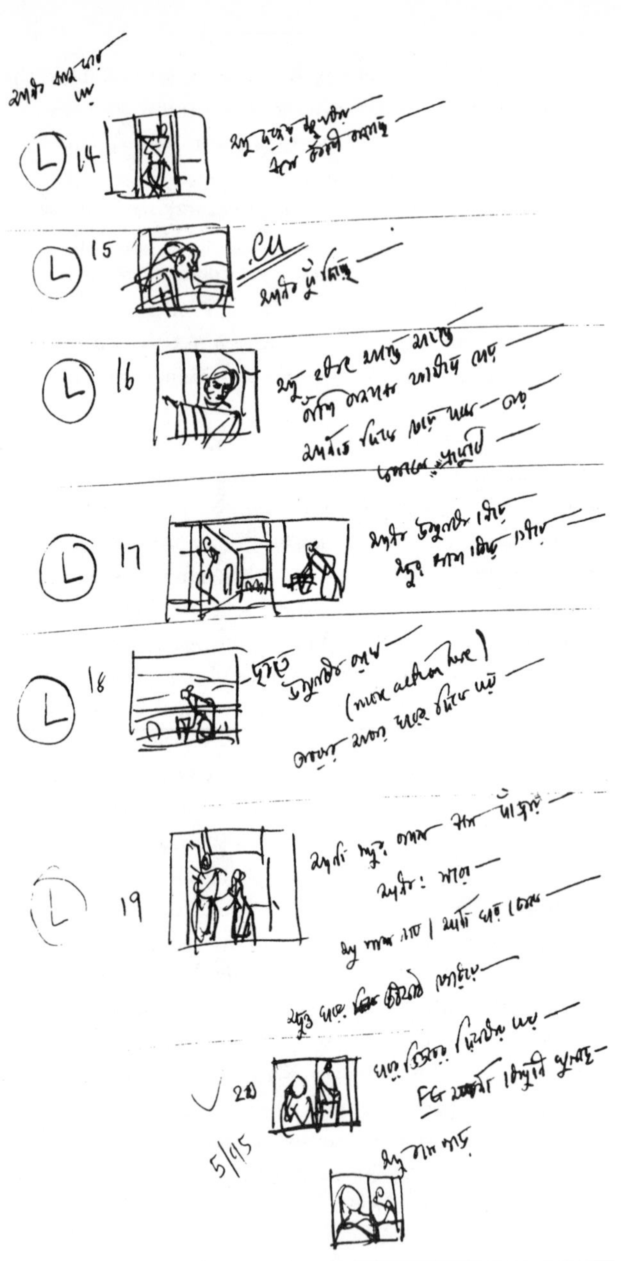

PAGE FROM SATYAJIT RAY SCRIPT FOR *Pather Panchali*
COURTESY CINÉMATHÈQUE FRANÇAISE

Most of the distributors visited by Ray had never seen a screenplay so complete in detail, nor a proposal so fully conceived and presented. Most hardly knew what to make of this material. It seemed impressive but irrelevant. They wanted to know a few simple things. Who were the stars? Who was writing the songs? Where were the dances? When Ray explained he had a different kind of film in mind, most concluded that it would not be a good risk. More than thirty distributors said "no."[4]

In some cases the "no" was not absolute. At Aurora Film Corporation, a forty-year-old company stemming from tent-show beginnings, managing director Ajit Bose was fascinated by the earnest visitor of towering physique with the vividly illustrated script. Bose said that he believed in the script and that Aurora would finance the film. Of course a professional director would be selected by Aurora to take charge. It was perhaps not an extraordinary stipulation to make, in dealing with a young man without a single screen credit. But to Ray the stipulation was out of the question.

Some gave him advice. They said his plans for location production were not practical, that everything he had in mind could be done better in a studio. In the Indian film world of 1952–53 virtually everything was done in a studio. The sort of thing Ray had in mind seemed to many a reversion to silent film days. Studio technology, the film industry was convinced, could accomplish almost any essential effect. On a few occasions Ray was goaded into passionate defense of his plans. Told that the rain scenes simply could not be done in the rain but required a well-equipped studio, he went into the monsoon rains with a 16-mm. Bolex for test sequences.

One day he found just the person he needed for the old aunt. She was a toothless hag who had once been a handsome and popular stage actress. Her career had been ended long ago by meningitis. Now permanently and painfully stooped, she lived from day to day with the help of opium tablets. She would be glad to play in the film, if it could mean the expensive tablets would keep coming. In a wavery but insistent voice she asked Ray: "Can you pay me twenty rupees a day?" Ray promised he would do so.

[4] Interview, Ray.

Having found her, he dared not delay the start of production. He assembled his production team. From his advertising agency earnings—substantial by Indian standards—every possible rupee began to go into weekend and holiday shooting. He sold his art books and phonograph records to help meet production costs. In a few months almost Rs. 20,000 of personal funds had gone into production. The film had hardly begun.

On the basis of the footage shot, a distributor now decided to risk an investment. A distribution contract was signed. The sum of Rs. 20,000 was made available for a continuation of the shooting. Weeks later the distributor looked at what had been done and changed his mind. He backed out and the work ground to a halt.

At some time during these months of rising and falling fortunes, the American director John Huston came to Calcutta, met with members of the Calcutta Film Society, talked with Ray, and saw sequences of *Pather Panchali.* His enthusiasm spurred new hope and determination and also had other effects. John Huston mentioned the film to Monroe Wheeler of the Museum of Modern Art, New York, who was in India planning an exhibit on the arts of India. Wheeler became interested.

About this time Ray, in desperation, approached the government of the State of West Bengal for funds for *Pather Panchali.* There was no precedent for help from the state. But some state officials knew of the interest of the New York museum official, and this may have played a part in the final decision, which was made at the highest levels of state government. The film became a state project under the aegis of its director of publicity. West Bengal put up Rs. 200,000 for the completion of the film. The state became its owner and would, henceforth, call itself "producer" of the film, although its participation was solely financial.

Production was resumed in earnest. But because Ray was still working for the Keymer agency, progress was slow.

In 1954, when the film was nearly finished after two years of work, an invitation came to have the film premiered at the Museum of Modern Art, New York, on the occasion of its Indian exhibit. The deadline necessitated a supreme effort to bring the film to a finish. Days and nights of intensive editing followed. Ravi Shankar completed his brilliant musical score in a matter of hours. There was a series of all-night recording and mixing sessions. Reels were rushed to and from laboratories. When Satyajit Ray finally took his package of film cans to the Calcutta air-freight office of Pan-American, and stood at the counter awaiting his turn, he fell asleep leaning on his package.

The world premiere at the Museum of Modern Art had some long-range results. Edward Harrison, who had scored successes in the distribution of Japanese films in the United States, was not present at the showing but, hearing about the film, arranged for a private screening a few days later. This led to eventual American distribution by Harrison not only of *Pather Panchali* but also of later Satyajit Ray films. The groundwork for this development was actually laid before Indian distribution had begun.

The State of West Bengal, as owner of the film, meanwhile entrusted its Indian distribution to Aurora Films, which thus once more came into contact with the young man with the script. The film, which was in the Bengali language, opened early in 1956 at three Calcutta theatres simultaneously. It started slowly, at first puzzling many patrons. Then it took hold, began to draw crowded theatres, and ran thirteen weeks. In the Bengali home market Ray had scored a substantial success. In the rural areas of West Bengal the film was likewise successful, but on a smaller scale. It stayed two weeks at many smaller theatres.

It must be remembered that after the India-Pakistan partition a Bengali film could be understood by less than 10 percent of the people of India. In India a Bengali film, even a hit film, had a normal theatrical distribution in West Bengal only. After that its Indian career was practically over. It might have some Pakistan distribution, but this market was becoming progressively more restricted and less profitable. Within India a Bengali film of unusual interest might also have special "Sunday morning showings" at theatres in New Delhi, Bombay, or Madras, for small Bengali-speaking colonies in those metropolitan centers. Such showings had prestige value but yielded little income. Dubbings into other Indian languages were almost never made. Extraordinary success might lead to the making of new versions in other Indian languages, but

Pather Panchali was so completely of Bengal that such action seemed implausible in this case. Subtitled versions were not generally considered practical, in view of limited Indian literacy.

Thus successful exposure to the Bengali-speaking people of India would normally have ended the career of *Pather Panchali,* even within India. The rest of India might not even have become aware of it or of Satyajit Ray, had it not been for the Cannes festival.

At Cannes the West Bengal entry, directed by an unknown, was at first not taken seriously by the festival management. At one phase of the program planning it was assigned to a morning showing, which would mean that only a handful of people would see it, while some of the jurors still rested in bed. The "important" films were supposed to come in the late afternoons or evenings. But a handful of people in Cannes—they included Edward Harrison and several others—had seen *Pather Panchali* and regarded it as "important." After crucial backstage struggles *Pather Panchali* was rescheduled for an afternoon showing immediately after a film by the Japanese director Kurosawa. But the Japanese delegation had arranged a large party after the Kurosawa film and some of the judges adjourned for this important social occasion. Next day the French critic André Bazin journalistically protested these events as "the scandal" of the festival and his protests led to a rescreening of *Pather Panchali.* Finally assembled, the judges were astonished at the Indian film and voted it the "best human document" of the festival. Thus began a sequence of awards which was to make *Pather Panchali* known on every continent, placing Ray almost at once among the great directors of the world and launching an extraordinary career. In the following twenty-five years he would direct some twenty-five films of superb craftsmanship, for many of which he would also write the script and compose the music. Almost all would earn back their production cost within Bengal, and most would also win foreign showings. Many would win international festival awards and keep his name almost constantly in the forefront among major directors.

Almost all would be in the Bengali language, so that only Bengalis would experience them in their native tongue. Although the name of Ray would become widely known in India as a result of his international fame, his films would tragically remain a closed book to millions speaking only Hindi, Telugu, Tamil, Marathi, Gujarati, Kannada, Malayalam, Assamese, Oriya, Kashmiri, and other languages. After some years the fame of Ray would lead to *evening* theatrical showings of several of his films in New Delhi, Bombay, and Madras, mainly in the form of "trilogy festivals" or "Satyajit Ray festivals." But being in effect foreign-language events, these would attract a somewhat specialized following, comparable to art-theatre audiences in some other countries. In a few instances the films of Satyajit Ray would be run with English subtitles—among the comparatively rare uses of this technique in India. But by and large Satyajit Ray would remain a Bengali film maker. In one sense this would be a limitation; in another, the key to his growing stature.

From the success of *Pather Panchali* in the Bengali film markets of India the West Bengal government earned back twice its Rs. 200,000 investment. Meanwhile it also received, during the first five years after release of the film, the following sums from foreign distribution:

United States	Rs. 251,230
China, People's Republic	40,000
Germany, France, Austria	25,000
United Kingdom	18,642
Poland	13,337
Thailand	11,880
Ceylon	6,612
Iran and Persian Gulf	6,500
Mexico	2,700
Netherlands	1,006
Total	Rs. 376,907

Along with receipts from Indian distribution, the state's income had thus reached, in these years, a total of about Rs. 800,000. And earnings were continuing.[5]

There are several astonishing elements in the success, financial and artistic, of *Pather Panchali*. Satyajit Ray, directing his first film,

[5] Interview and correspondence, P. S. Mathur.

had decided not to use any of the established cameramen of the Indian film industry. Feeling that all were too saturated in pictorial formulas of the industry, he had gone instead to a much-admired still photographer, Subrata Mitra, and invited him to shoot his first motion picture. *Pather Panchali* was thus written, directed, and photographed by "newcomers."

Comparable to the selection of Mitra was that of Ravi Shankar, who composed and played the score. He was a musician of enormous celebrity, but not associated with "film music." Among the members of the cast, those with acting experience included the old aunt, the father, the rich lady next door, and the schoolteacher. Most of the performers were without professional experience.

A certain number of technicians with film backgrounds were enlisted in the enterprise. These included the film editor and the art director. The latter played a crucial role. Although outdoor scenes were shot at locations a dozen miles from Calcutta, much of the indoor work was finally done in a Calcutta studio, and sets for this were designed and built to match location structures. In using this procedure, Ray had to some extent adjusted his original plans, but without compromise to results. No hint of "studio quality," in scenic texture or lighting, was permitted.

The group worked with relatively simple means but with great technical resourcefulness. Ray had an intense dislike of "slick" light effects and became devoted to "bounce lighting," with the light directed at one enormous reflector.

A memorable sequence of *Pather Panchali* showed the young girl, Durga, running through the woods. Keeping her in close-up, sharply in focus, the camera appears to move along beside her throughout the sequence. The effect would normally be gained by a trucking shot in which a camera moves on a track parallel to the running girl. No such equipment was available to the group. In actuality the sequence was shot by a stationary camera. The girl ran through the woods in an exact circle around this camera, which panned to follow her, using a telephoto lens. To ensure perfect focus, her course through the woods had been laid out by measurements with a piece of string from the camera position. The course ended where it began. No trucking shot could have been more precise in effect.

Most of *Pather Panchali* was postsynchronized; dialogue recorded on location was not actually used except as a guide in the recording of the final sound track, made under controlled acoustical conditions. Ray here followed the method of the Italian neorealists.

Satyajit Ray's prominent use of artists and technicians who were newcomers, at least to film, represented in some respects a repudiation of the Indian film industry and its prevailing tenets. The long failure of Indian films to win recognition in Western markets had generally been ascribed by industry leaders to insufficiency of technical resources. Occasionally they had journeyed to Hollywood in search of "know-how." They came back in awe of equipment they had seen. They duplicated Hollywood technical devices as best they could. If they only had greater resources, what they could not do! When finally an Indian won success on Western screens, it was achieved not by lavish equipment or vast resources. Though the makers of the success had a proper respect for technical precision, their victory had been won by something else: primarily, by integrity in the handling of content.

Not surprisingly, the victories of Ray aroused mixed feelings in various quarters, both in film industry and in government. The industry hailed the Satyajit Ray successes. But an undercurrent of pique was evident, especially in Bombay and Madras. In these centers it became customary to say that Ray's films were of course splendid "artistically," but that Bombay films—or Madras films—were better "technically." Evidence that Ray's films are considered, in Europe and America, superb on both technical and artistic grounds has not seemed especially welcome. There has of course also been a feeling of discomfiture about the successes of a project rejected by numerous private financiers and eventually financed by a state government.

In the central government official jubilation also had, from the beginning, some contrary undercurrents. There was of course delight that Ray had "put India on the international film map." He received a presidential award and other honors. But the success of *Pather Panchali* demonstrated to some the shortsightedness of central government policies toward film. In 1954, after three years of silence, the government had put aside the idea of a Film Finance Corporation, as recommended by the Film Enquiry Committee. The central government considered the plan impractical for financial reasons. Now a state government had, in startling fashion, dramatized the arguments that had been advanced for the idea. It had shown that an alternative source of capital, controlled by a different set of values, could indeed liberate a film maker from success formulas dominating an industry. The state had even demonstrated that such government use of capital need not necessarily result in loss but might yield profit. It is not surprising that the central government now changed its mind about the feasibility of the idea. In 1957 the Ministry of Information and Broadcasting announced, almost as though it were a new idea, the government intention "to set up a Film Finance Corporation for the purpose of rendering assistance to film producers by way of loans."[6] Parliamentary approval was given in 1959 and the agency finally began operation in 1960, eight years after the recommendation of the Film Enquiry Committee—and thirty-two years after a similar recommendation of the Indian Cinematograph Committee. The West Bengal success appears to have had a part in spurring the action.

While the central government officially rejoiced over the success of *Pather Panchali,* some highly placed officials appear to have frowned on the film and especially its distribution abroad. The objection was that it pictured India in terms of poverty and that this damaged India's international image. This opinion never achieved the upper hand in decisions relating to *Pather Panchali,* but the persistence of the attitude was evident in various developments. In September of 1956 the magazine *Filmindia* reported:

> The Government of India has directed that before any State Government sends films—features or documentaries—abroad for exhibition, the State Government should ascertain the film's suitability from the point of view of external publicity by the External Affairs.
>
> Recently a State Government made direct arrangements for showing its film abroad with a foreign distributor, by-passing the External Publicity Division.[7]

[6] *Report, 1956–57,* p. 36. [7] *Filmindia,* September, 1956.

SATYAJIT RAY'S *Pather Panchali,* RELEASED 1956

The *BMPA Journal* added a detail. An Indian mission abroad had commented on the film's "unsuitability from the point of view of external publicity."[8] Government spokesmen, while not confirming that all this related to *Pather Panchali,* would not identify the film involved.

Later the Central Board of Film Censors issued a new listing of material that might cause censorship; it included "abject or disgusting poverty." Again government spokesmen did not identify the film or films precipitating this regulation.[9] To any Satyajit Ray admirer it seemed inconceivable that the regulation could relate to work so rich in warmth and humanity, yet some in the central government appeared to regard it in that light.

But the more important fact was that Ray's debut as a film maker, in spite of undercurrents of hostility, had set changes in motion at

[8] *BMPA Journal,* August, 1956.
[9] *Annual Report 1959–60* (South Indian Film Chamber of Commerce), p. 2.

home as well as abroad. It was quickly evident that he was influencing younger Indian film makers—at first, mainly in Bengal. Ray's own position was also undergoing important changes. Backers who had shied at Satyajit Ray's *Pather Panchali* proposals were ready to finance subsequent projects. Aurora Films, which had expressed willingness to finance *Pather Panchali* only in accordance with its own dictates, was ready to finance its sequel, *Aparajito* (The Unvanquished), under Ray's direction and control. It appeared in 1957, winning the Golden Lion award at Venice and the Selznick Golden Laurel trophy in the United States. A variety of private backers provided finance for most of the later features, including *Apur Sansar* (The World of Apu), which in 1959 completed the "Apu trilogy" and won further honors—in fact, a whole sequence of major festival awards.

The Apu trilogy formed an epic that moved from the village to the city, from the distant past to the recent past, from a structured rural life to an anarchic metropolitan milieu, with all its uncertainties for the future. This epic movement may have been especially meaningful to Satyajit Ray because it held parallels to the experience of the Ray clan, which had its roots in rural Bengal and had known many reversals of fortune, finally becoming a Calcutta family deep in the conflicts and problems of the modern world. It also represented the history of modern India. And it foreshadowed the scope of Ray's work in the decades to follow: its panoply of figures would eventually seem like figures in a vast landscape representing more than a century of Indian history.

Ray would explore this historic landscape in a diversity of ways and moods, and from many points of vantage, looking at it in turn through the eyes of the dispossessed, the landlord, the job-seeker, the potentate, the corporate executive, the peon, the devotee, the skeptic, the colonial administrator, the native aide, the revolutionary, the time-server.

A most important aspect of this enormous canvas may be summed up in a very simple statement by Ray himself: "Villains bore me." It was a statement that calmly repudiated much of India's film output—in fact, much of the world's film output—and for good reason. The villain is an age-old formula for evading and obscuring

SATYAJIT RAY'S *Aparajito* (THE UNVANQUISHED), 1957

historic truth. Ray, in many of his works, explored the complex of devices by which establishments have legitimized and bolstered their status, in ways that have won consensus among the ruled as well as the ruling, and that have often eluded, for much of the time, the consciousness of both.

Thus Ray examined many figures of Indian society representing power and privilege, and those who willingly or unwillingly accepted the dominance. He scrutinized the world of the zamindar, as in *Jalsaghar* (The Music Room), 1958; of the husband, as in *Charulata* (Charulata), 1964; of the movie star, as in *Nayak* (Hero), 1966; of the businessman, as in *Aranyer Din Ratri* (Days and Nights in the Forest), 1970, in *Pratidwandi* (The Adversary), 1970, and in *Seemabaddha* (Company Limited), 1971; of the Brahmin teacher, as in *Ashani Sanket* (Distant Thunder), 1973; of the colonial commander, as in *Shatranj Ke Khilari* (The Chess Players), 1977. All these figures have appeared in many works as heroes or villains, instantly recognizable as one or the other. In Ray's work the interest is in the psychology, the symbolism, the power structures that have sustained their hegemony. In this sense the films comprise a rich education in the workings of society.

Ray's integrity in the handling of social history has tended to alienate many. It upset comforting stereotypes. Many people would have preferred villains. Some have resented what they considered his too-cool detachment. They said he was not "committed."

If these critics meant he did not enunciate political doctrine, they were right: Ray was determined to leave audiences grappling with the problems and situations he had laid bare, and not to offer the sugar pill of formula solutions. But if they meant that he was not concerned, they were far from right. Ray appeared to invest in his work enormous emotion, under tight control. In an interview with the critic John Hughes, he agreed that "revolution is certainly implied" in a number of his films. "There's the sense that things cannot go on the way they are."[10]

Along with overall coherence in concerns, Ray's films have shown a diversity of moods, techniques, and genres that was scarcely foreshadowed by his impressive beginning. In addition to his fiction films—which have included slapstick comedy, satire, fantasy, tragedy—he produced documentaries, including the moving *Rabindranath Tagore* (Rabindranath Tagore),[11] narrated by Ray himself and using archive material along with reconstructions. Although most of his films were feature-length, he also made brilliantly deft short films such as the three parts of *Teen Kanya* (Three Daughters),[12] 1961, based on Tagore stories, and the language-less *Two,* a gem made for American television but still wholly of Bengal.[13] While most films were realistic, he used surrealist devices in *Pratidwandi* and simple animation in *Shatranj Ke Khilari.* In *Devi* he produced an Ibsen-like problem play on the power of superstition.

In the midst of work on his dark "city films" focusing on the busi-

[10] *Film Comment,* September–October, 1976.

[11] A skeletonized 20-minute version of the one-hour documentary was distributed to theatres by the Films Division as an "approved film."

[12] Released abroad as *Two Daughters,* with one story omitted for reasons of length.

[13] It appeared on an *Esso World Theatre* program devoted to India.

SATYAJIT RAY'S *Devi* (GODDESS), 1960

ness world, he produced the rollicking, phantasmagoric *Goopy Gyne Bagha Byne* (Goopy and Bagha), 1969, based on a fable by his grandfather Upendrakishore. Its creatures included demons, wizards, animals, and people, and its climax involved efforts to prevent a war. Cartoons by Ray joyfully set the mood. Its dances and lyrics had elements of the Indian song-and-dance genre, and its cast of characters suggested the mythological, but all in a different and refreshing context. Very successful in India, it has been considered—perhaps mistakenly—less translatable than Ray's more serious films. This attitude, applied to other comedies also, has tended to give Ray a more somber image outside Bengal than he has at home.

Meanwhile he moved in an entirely new direction. In 1961, while working on his first color film, *Kanchanjanga* (Kanchanjanga), he decided to revive the children's magazine *Sandesh,* which his grandfather had founded and his father had maintained to his final days.

Perhaps Ray's closeness to his own son Sandip—born during the early work on *Pather Panchali*—prompted the revival of the magazine. In any case it led to a stream of Satyajit Ray stories and drawings for the magazine, with eventual impact on the films. The magazine output resulted in two published collections of short stories, nine comic-detective novels, as well as science-fiction and horror stories—"all the sort of things I loved as a child." Columbia Pictures became interested in one of the science-fiction stories. Ray turned one of the detective novels into the joyful film *Sonar Kella* (The Golden Fortress), 1974, generally listed as a "children's film"—a term Ray disliked. (He said it was for children of all ages, 10 to 80). Translations of the books were made into Hindi, Marathi, Gujarati, Malayalam, and Oriya. Thus while Ray was continuing to produce on average one feature film per year, he was pursuing an extraordinary parallel career in print.

In his films, Ray has been constantly concerned with the "social identity" of the characters. He believed that people act and react to each other in ways that derive not only from their personal characteristics but from their existence in a particular place and time in a particular social context. This matter of social context is, of course, ignored in most Indian song-and-dance films, which have tended to exist in a vacuum. The extraordinary believability of Ray's characters comes from their being firmly rooted in a defined society—usually, in some aspect of Bengali life in the nineteenth or twentieth century. Paradoxically, the feeling of "universality" has stemmed from our recognition of this authentic localism.

The context has generally been suggested by significant details, gradually forming an enveloping world. The colonial realm of *Charulata*—based on a Tagore work, and representing a Bengali counterpart to Ibsen's *A Doll's House*—is first suggested by wonderfully evocative shots in which we see the young wife Charulata darting from window to window in her mansion, studying the movements of the outside world through slats of the window blinds, using opera glasses. We sense a caged bird, and other details add themselves to this image. Her movements allow us to see details of the house, and to sense its period and status. We also see details of the outside world—passersby, their activities, interactions, and vehicles.

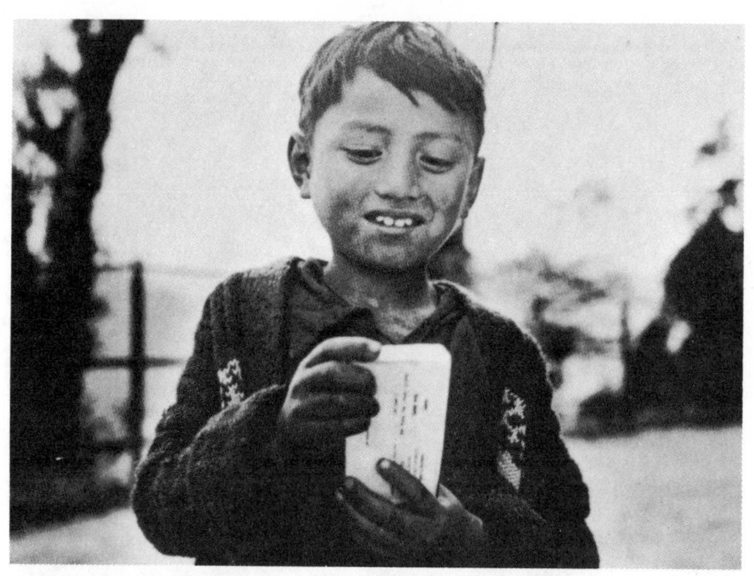

SATYAJIT RAY'S *Kanchanjanga* (KANCHANJANGA), 1962

SHOOTING *Kanchanjanga:* ON LOCATION IN DARJEELING, NOVEMBER, 1961. WITH SCRIPT, SATYAJIT RAY

SATYAJIT RAY'S *Charulata,* 1964

Especially we notice, along with her fastidious beauty, her eager curiosity about that outside world, a tremulous desire to know it more closely. Later we see her move to the interior gallery of the house, where her handsome, intellectual-looking husband, deeply engrossed in a book, is seen slowly walking past her without being aware of her close presence. She watches him as he walks to the far end of the gallery, where he stands reading. She raises her opera glasses and looks at him again—this time, as though seeing him too as of another world. Via such suggestions a Charulata world takes shape around her. Few film makers have matched Ray in this building of evocative detail.[14]

After two decades of Bengali-language features Satyajit Ray for the first time ventured a film in a language and culture not his own —*Shatranj Ke Khilari* (The Chess Players), released in 1977, dealing with an episode in the British take-over of India and set in the aris-

[14] The problem of "social identity" is discussed by Ray himself in his book *Our Films, Their Films,* p. 12.

tocratic, feudal Lucknow of 1856. The English characters speak English, the others Urdu, the Hindi-related language of the area—a language Ray did not know. The film was based on a story by Prem Chand.

In view of Ray's obsessive interest in "social identity," his long reluctance to undertake such a project was understandable. He was aware of miscalculations by other directors working under such circumstances. But he had often been offered backing for a film aimed at the Hindi market, and when the Hindi story by Prem Chand, dealing with a period in which he had long been interested, came to his attention, he felt impelled to make the plunge.

The story fascinated him because of its social implications as well as its cinematic possibilities. He decided to interweave two threads. One centered on General Outram of the East India Company ("The Company") preparing to annex the Kingdom of Oudh, the last quasi-independent realm of India, nominally ruled by Wajid Ali Shah. Major chunks of his kingdom had already been sliced away by the British, and the final take-over was now imminent.

Characteristically, Ray made neither General Outram nor Wajid Ali Shah the scapegoat of events. His General Outram is troubled over the illegality of the procedures he must follow, which require him to abrogate a "treaty of friendship" while forcing the king's abdication. But Outram is also a firm believer in the glory and destiny of empire, and regards it as his duty to history—and the Queen—to ignore his own scruples. Besides, he considers Wajid Ali Shah a weak and ineffectual king—which he is. Ray's attitude toward the king is similarly qualified. He depicts him as an accomplished poet and musician, with no great enthusiasm for the maneuverings of statecraft, who has relied on his "treaty of friendship" with Britain to pursue his cultural interests. When Britain's treachery becomes clear, the king is at first determined to resist with force, but finally decides, by surrendering, to avoid shedding the blood of his people.

Interwoven with these events are episodes involving the noblemen Mir and Mirza, who have an unrestrained passion for chess. The arrangements under which the Kingdom of Oudh has been allowed to remain ostensibly independent, taxing its people while receiving "protection" from the British, have enabled these noblemen to continue their aristocratic life undisturbed, endlessly playing chess by ancient Indian rules, oblivious to the rougher chess being played by the British. The men are trying to save their kings as their King goes under. They are even oblivious to the needs of their own wives, one of whom is bedding down with her husband's nephew.

Thus Ray's portrait of the fall of Oudh is a sardonic one. Dissecting colonialism in a way that is equally applicable to neocolonialism, it seemed to some observers to be extraordinarily timely. But if it did not at once enthrall Hindi film audiences, reasons are not hard to find. Indians could not readily derive emotional satisfaction from its version of history—nor, for that matter, could the British. If it had no villains, it had no heroes either—and no action climax, such as many may have yearned for.

If there was a message, it had to do with non-involvement—with playing chess while Oudh fell. It was perhaps significant that the film appeared during a Kafkaesque period of recent Indian history, the period of "Emergency," likewise made possible by massive non-involvement. This may help to explain why *Shatranj Ke Khilari* won critical acclaim but not immediate box office success. One critic called it a film "not for the heart but for the head."[15] It may, in the long run, be considered one of Ray's truly important films.

Most of Ray's early features were produced on budgets of less than Rs. 500,000—considered shoestring financing in the Hindi film world but not unusual in Calcutta's truncated market, where a "general air of privation" continued to pervade studios and laboratories even as Ray grew famous.[16] *Shatranj Ke Khilari,* with its cast of well-known professionals, cost a good deal more—some Rs. 2,000,000—but this was still modest by the standards of Bombay, where Rs. 4,000,000 had come to be considered normal and Rs. 10,000,000 not unreasonable. More than half of this might go to the stars.

[15] Khanna, "Reviews," *CSSEAS Review,* Vol. I, No. 2.
[16] Ray, "Problems of a Bengal Film Maker," in *International Film Annual,* No. 2, p. 51.

Throughout his first twenty-five years—twenty-five features—of film production, Ray's working group remained surprisingly constant. He had the same editor throughout the period, and only two cameramen. One assistant left him but later returned. This continuity involved no contracts, no permanent organization. Members of the team simply endeavored, again and again, to be available when Satyajit Ray was ready for his next picture. Partly out of a sense of obligation to them, he moved from one production to another at a rapid pace. At all times he kept various possible projects at various stages of preparation. Screen rights to a number of works were discussed with authors and publishers. Complete adaptations of public-domain classics were written, then held in abeyance. Projects sometimes waited for precisely the right actor or actress—or for a budget of the right magnitude. To the question—What will Satyajit Ray do next?—there were always several possible answers.

The boy Sandip gradually became one of Satyajit Ray's most valued assistants—"the best assistant I ever had"[17]—suggesting that the close family collaboration of past decades was continuing into a fourth generation. Satyajit had long treated Sandip as an adult, which seems to have been a Ray family tradition in the treatment of children—a key, perhaps, to the success of the children's magazine.[18]

During his first decade of film making Ray seemed a towering but solitary figure. Some thought they saw a "movement" following in his steps. Figures of unquestioned talent won attention on the Bengali scene.

There was Tapan Sinha—like Ray, an alumnus of the Calcutta Film Society. His direction of several Tagore stories— *Kabuliwalla* (The Merchant from Kabul), 1956; *Kshudita Pashan* (The Hungry Stones), 1960; *Atithi* (The Runaway), 1966—showed a controlled, sensitive style similar to that of Ray, as did his direction of *Hansuli Banker Upakatha* (Folk Tales of the River Bend), 1962.

There was Ritwik Ghatak, whose work suggested unpredictable genius. His *Ajaantrik* (Pathetic Fallacy), an ebullient fable of 1958

[17] Interview, Ray.
[18] Seton, *Portrait of a Director,* pp. 273–75.

about a taxi-driver's love for his aging taxi—and the taxi's response—won him considerable vogue. But his experience as a refugee from East Pakistan haunted much of his work, and led to somber if brilliant epics such as *Meghe Dakha Tara* (The Cloud-Capped Star), 1960, one of several works reflecting his refugee-camp days.

There was Mrinal Sen, another who found inspiration in the Calcutta Film Society—when he could afford to go. His first fling at direction, *Raat Bhore* (Night's End), 1956, was so disastrous that he determined to renounce all film making. But a few years later his *Neel Akasher Neechey* (Under the Blue Sky), 1959—a film of Maoist influence—won him commercial success and a following which tended to prefer him to Ray as being more "committed." His 1960 *Baishey Shravana* (The Wedding Day) won him a showing at Venice, and a promising career appeared to await him.

There was also Rajen Tarafdar, whose film *Ganga* (Ganges), was shown at the 1961 Venice festival and was praised by *Variety* for its "robust approach to reality."[19]

And there were others. But these were of Bengal, and this helped further the impression that they represented a Calcutta aberration, with no significance for the larger Indian film world. Until the late 1960s there was little evidence that Ray, or the handful of other seekers of alternative forms of cinema, could have a wide impact on the industry or its audience. But then, slowly but surely, changes seemed to take hold.

Turning tide

The changes that came over the Indian film world in the late 1960s and the 1970s had many causes, with diverse threads of influence.

Some were international. Indian film makers had increasing contact with foreign film artists and their work. This had begun in the 1950s, when the First International Film Festival of India—held in 1952—drew a number of distinguished foreign film directors to Indian film centers. The next International Film Festival of India took place in 1961 and was followed by others at shorter intervals,

[19] *Variety*, September 13, 1961.

RITWIK GHATAK'S *Ajaantrik* (PATHETIC FALLACY), 1958

TAPAN SINHA'S *Hansuli Banker Upakatha* (FOLK TALES OF THE RIVER BEND), 1962

INDIAN DELEGATION IN THE USSR, 1951. LEADING APPLAUSE, PUDOVKIN. IN WHITE, SUBRAHMANYAM. EXTREME LEFT, COMEDIAN KRISHNAN AND HIS WIFE

becoming annual events in the 1970s. All increased the international film traffic, which also moved in the opposite direction, as more and more Indian film artists and their works were invited to foreign festivals, resulting in an inevitable cross-fertilization.

The traffic was to some extent spurred by diplomatic rivalries. A group of Indian film artists was invited to Moscow in 1951 and showered with hospitality. Each later received, in 35mm film, a documentary record of his or her visit, produced by Mosfilm, which some of the artists released to theatres. The following year another Indian group made a United States tour, which included a White House visit with President Truman and a lavish Hollywood banquet. In later years such intercultural expeditions became more frequent and they sometimes generated international collaborations of various kinds.

Such ventures always put international amity to a severe test. This was the case with *India '57,* an Indo-Italian project under

GREETINGS FROM TRUMAN: INDIAN DELEGATION AT THE WHITE HOUSE, 1952. AT TRUMAN'S LEFT, NARGIS. JUST BEHIND HER, RAJ KAPOOR

INDIAN DELEGATION IN HOLLYWOOD, 1952. CHANDULAL SHAH GETS A KISS FROM SUSAN CABOT. AT SHAH'S SIDE, MILTON RACKMIL

Roberto Rossellini, launched with euphoria but terminated when his visa was not renewed because of a widely reported amorous scandal.[1] More successful was *Pardesi* (The Traveler), a 1957 Indo-Russian venture organized by the ubiquitous K. A. Abbas—involving Mosfilm and widely distributed in both India and the Soviet Union, which was said to have ordered 600 prints.[2] A later Indo-American project, *Nine Hours to Rama,* launched in 1962 with rupee funds of Twentieth Century-Fox, was another misadventure, resulting in a film rejected by Indian censors.[3] More fruitful was the Indo-American collaboration of Ismail Merchant and James Ivory, whose 1963 film *The Householder* had a score by Satyajit Ray and featured young Shashi Kapoor, brother of the star Raj Kapoor. It won Columbia Pictures as distributor abroad, providing funds for a 1965 Merchant-Ivory release, *Shakespearewallah* (Shakespeare Peddler), with Shashi Kapoor and Felicity Kendall. It re-enacted adventures of the Kendall family, which had long toured India with Shakespeare performances, building a company that had included Shashi Kapoor, and in which he met his future wife, Jennifer Kendall. The production, which involved various members of the Kendall family, won critical acclaim—while helping to propel Shashi Kapoor toward superstardom, and paving the way for further Merchant-Ivory intercultural productions. Projects of this sort, successful or not, inevitably yielded increasing awareness of the techniques and ways of others, and could not fail to influence ideas and aspirations.

During these years a more pervasive influence was the film society movement. At the time when Satyajit Ray had begun his film career, only a handful of film societies existed in India, and some had collapsed after a short life. Burdened with various taxes, they had also faced constant procedural hurdles such as negotiations over import licenses and censor certificates—harassments from which non-profit film societies in many countries had been relieved in recognition of their educational role. An Indian film society could at first

[1] *Lok Sabha Debates,* December 18, 1957.
[2] Its genesis is described in Abbas, *I Am Not an Island.*
[3] For the disputes generated by the production see *Blitz,* January 20, 1962; and *Screen,* January 26, February 2, 1962.

sidestep these entanglements only by holding screenings at foreign embassies, many of which were eager to invite them and to provide films. But this tended to make the society a transmission belt for propaganda, subtle or otherwise.

In the 1960s the hurdles began to be lowered, partly in consequence of the prestige of Satyajit Ray. In 1962 a new censorship arrangement was worked out. Censorship would be waived for a film if Ray, as president of the Federation of Film Societies of India (FFSI), signed a statement affirming that he had seen the film and considered it suitable for film societies. The arrangement had obvious drawbacks: Ray could hardly see enough films to serve the needs of the movement, especially after it began to grow. There was also the problem of coping with diverse languages. The arrangement was soon altered by the establishment of a Preview Committee representing the FFSI. Any one of four regional vice presidents would now report to Ray that the film was considered suitable by committee members listed in the report, and Ray would sign the waiver on this basis.

An effective voice in furthering the movement and laying the groundwork for these arrangements was Marie Seton, film historian and biographer of Satyajit Ray. She also helped to win the interest and support of Indira Gandhi, who became vice president of the FFSI and continued in this position even after she became, in 1964, Minister of Information and Broadcasting. The movement was meanwhile growing by leaps and bounds, aided in a number of states by exemption from entertainment taxes. By the 1970s 150 film societies were affiliated with the Federation and scores of others were asking for affiliation. Total membership had risen to some 70,000–80,000. One of the largest societies, with some 2,000 members, was the New Delhi Film Society, attended in its early years by Indira Gandhi. Its location helped to make it an effective influence on the attitudes of government officialdom.

Two film societies in Bombay, the Anandam Film Society and the Film Forum—long presided over by K. A. Abbas—played a significant role in the education of future film makers. A member of both was Basu Chatterji, who was a film society devotee for years before it occurred to him to consider a film career. A cartoonist for *Blitz,* he became especially enchanted with comedies emanating from Czechoslovakia, directed by Milos Forman, Jiri Menzel, and others. But it was the films of Satyajit Ray, also regularly included in film society programs, that finally confronted him with the thought: "We too can do such things." And he began to make plans. Other Bombay film society devotees were Basu Bhattacharya, who became a film maker shortly before Basu Chatterji, and Shyam Benegal, who, long before turning to film making, had been one of the founders of the Hyderabad Film Society. In Kerala a number of future film makers, including G. Aravindan, obtained their first tutelage from film societies.

The societies had critics as well as evangelical supporters. One accusation was that people joined to see uncensored "sexy" films. Some film society leaders worried about the accusation and urged a careful screening of applicants to weed out the dubiously motivated, but others rejected this course. One insisted that people might join "for the wrong reasons" but stay "for the right reasons." By and large, devotees rejected the call for exclusivity. K. A. Abbas was among those who campaigned for a "wide open" policy.[4]

While film societies were providing a many-sided orientation—for audiences, officials, and film aspirants—more specific training was becoming available at the Film Institute of India, which began operation in 1961, after the government purchased for its use the studio lot in Poona that had once been the home of Prabhat, but had long stood empty. Here students acquired huge studios and screening and editing rooms, along with a woodland area and a small lake with diverse shorefronts, all created by Prabhat. It soon attracted students from all states and from other countries of Asia and Africa.

The Institute increased in significance when joined in 1964 by an important new creation—the National Film Archive of India. It amassed an international assortment of film classics and an extensive collection of Indian films, some rescued from old vaults, closets, and attics. They ranged from pre-Phalke "topicals" and surviving Phalke treasures to major films of later decades. Under the leader-

[4] Interviews, Vinod Mehra, H. B. Mathur, Kalyan Banerjee, I. S. K. Devarayalu.

ship of P. K. Nair, the archive began to play a role similar to that of France's *Cinémathèque* under Henri Langlois. To many film students the daily screenings of archive classics became a key element in their training, a stimulus to restless discussion and experiment. In 1974 the Institute acquired a television wing and became the Film and Television Institute of India, further broadening its potential.

As educational forces, Institute and Archive interacted with the film societies and festivals. Leading film makers visiting Indian film festivals also regularly visited film societies and the Institute. Film society pioneer Satish Bahadur, a founder of the Agra Film Society, became a leading member of the Institute faculty, often presiding over the screenings of classics and visits by film makers, and providing a historical context. The Archive acquired a selection of classics available for loan to film societies.

Amid all these converging influences, future film makers were growing up in a rapidly changing environment. When the Institute met its first students in 1961–62, they had seemed to be aware mainly of the latest hits, the stars of the moment. Students' expectations and aspirations had been shaped by these. Before long they began to see themselves as part of an evolving and dynamic process, with an international heritage and unguessed possibilities for the future. They were aware also of the continuing phenomenon of Satyajit Ray. Yet the result of these cumulative influences seemed in doubt until the late 1960s, when a series of explosions began. The spark seems to have been lit by the Film Finance Corporation.

The FFC, dating from 1960, had begun cautiously. One of its first loans was to the veteran V. Shantaram for a new version of the ancient Shakuntala story, released in 1962 as *Stree* (Woman). It had its quota of musical numbers and could probably—it was widely believed—have secured private finance. Perhaps the FFC, at this time chaired by a former income tax commissioner, wanted to establish a business-like record. At any rate, early loans tended to follow this cautious pattern—not always with financial success, which made it seem worse.

The loans assisted some projects of unquestioned stature, including three Satyajit Ray films that were among his finest—*Charulata,*

Nayak, and *Goopy Ghyne Bagha Byne.* But even in these instances, the FFC was criticized for aiding the already successful.

In the late 1960s, under the chairmanship of journalist B. K. Karanjia, the FFC shifted its policy to favor low-budget, experimental proposals. It was aware of the risks involved in this shift. In view of the obsession of exhibitors with star names, films financed in this way could easily be ignored by distributors, and become shelf items. The shortage of theatres gave exhibitors a bargaining power dangerous to the new policy. In fact, a number of the experimental films did remain undistributed, bringing predictions that the FFC would be dissolved.

But in 1968 the FFC suddenly began a series of dramatic successes. These started with *Bhuvan Shome* (Bhuvan Shome)—a pivotal film in Indian film history—directed by Mrinal Sen and featuring playwright-actor Utpal Dutt and the unknown Suhasini Mulay. Filmed on location with a small cast, it was simple in structure, rich in resonance. Produced in Hindi (though by a Bengali film maker), it was a total departure from Hindi formulas, yet it scored box office successes while also winning the year's Best Feature Film, Best Actor, and Best Director awards, along with international honors. In one blow, the FFC had apparently freed itself of the jinx attached to its backing.

Bhuvan Shome was indeed a happy conjunction of spirited talents. Those in the audience favoring "committed" films could find many meanings in it, but it wore its meanings lightly. Shome, a railroad executive who is a terror to all his employees, goes bird-hunting in the country, where he becomes dependent on a farm-girl who is not afraid of him, and even looks on him with indulgent amusement—an experience new and jolting to him, with an eventual humanizing impact. The fact that the unknown Suhasini Mulay, playing her first film role, was in no way intimidated by the prestigious playwright Utpal Dutt and could treat him with precisely the amused condescension the role required, produced a give-and-take full of sparkling, semi-improvised exchanges.

Bhuvan Shome gave a new impetus to the career of Mrinal Sen, who followed with further vigorous works, mostly backed by private finance. All his films had political ramifications, sometimes explicit,

MRINAL SEN'S *Bhuvan Shome* (BHUVAN SHOME), 1969:
INTRODUCING SUHASINI MULAY

MRINAL SEN

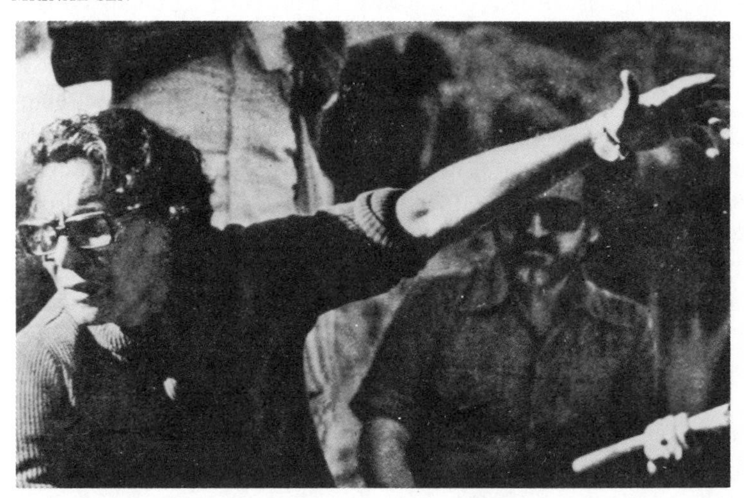

as in the angry *Calcutta '71,* sometimes muted yet unmistakable, as in *The Interview.* The latter takes an ebullient Brechtian approach: the audience is told, and at various times reminded, that the hero who is looking for a job in Calcutta actually *is* looking for a job in Calcutta, and that his efforts are meanwhile being filmed by his friend Mrinal Sen, the film maker. When the hero rides on a crowded bus, we glimpse a cameraman among the swaying riders. The plot revolves around an impending interview with a multinational corporation, requiring a Western suit, not Indian garb. The hero has taken his one suit to the cleaners, but the cleaners pick the fateful day to go on strike. The resulting disaster makes mordant comment on the impact of multinationals on the struggling world of Calcutta. There is anger, but it is laced with irony and humor.

In *Chorus,* 1974, Sen again tackled the Westernized corporation, this time in a surrealistic vein. The corporation is housed in a castle, with all the accouterments of medieval power, and a job availability announcement brings a line of countless thousands to the castle gate, producing an eventual revolution. The device had extraordinary visual impact at the start, but was difficult to sustain. The film reflected the oft-repeated view that multinationals had moved into the imperial position once occupied by the British. It was no surprise that in a subsequent film, *Mrigaya* (The Royal Hunt), Sen harked back to the British raj, centering on its impact on a tribal community. With this film he again won—in 1976—the year's Best Feature Film award.

Mrinal Sen, unlike Satyajit Ray, seemed ready to work without inhibition in diverse languages not his own. He did so with success in Oriya in the 1967 *Matira Manisha* (Two Brothers); in Hindi in *Bhuvan Shome* and the 1972 *Ek Adhuri Kahani* (An Unfinished Story); in Telugu in the 1977 *Oka Oorie Katha* (A Village Story). This adaptability tended to give him a national name. His ability to make intelligently political films palatable to audiences made him a novel phenomenon in Indian films. He seemed to lead the way in a new political trend in film.

The FFC's success with *Bhuvan Shome* was followed by others of perhaps greater significance in the nurture of new talent. Within a few years it financed first films for a number of new directors, launching several remarkable careers. First came Basu Chatterji with *Sara Akaash* (The Whole Sky), 1969—his first attempt at both scriptwriting and directing. With photography by K. K. Mahajan, an alumnus of the Film Institute of India, it was a new-talent showcase of some importance. Chatterji's sole preparation, outside of attendance at film societies and the reading of books, was to serve as assistant to his friend Basu Bhattacharya during the previous year, when Bhattacharya was making his first film.[5]

Sara Akaash was a low-budget film of exceptional simplicity, made without stars. It concerned two young people who get married via arrangements made by their parents. Strangers, they are scarcely able to talk to each other for two months, but finally the ice is broken. Chatterji told the story in small, deft episodes, whose poignancy and quiet humor were reminiscent of the Czechoslovakian films he had admired at film society showings. Their mood and simplicity departed sharply from Bombay tradition, yet the film won instant popularity and the Best Screenplay award, and led to a stream of Basu Chatterji successes that kept his name on theatre marquees for years—before those of performers. "A Basu Chatterji film" became the selling phrase. Private finance was now readily available.

Chatterji continued to exploit the vein he had pre-empted with *Sara Akaash*. In almost all his films he focused on a young man and a young woman—the subject matter of most Hindi films, but with a difference. In his films they did not meet in extravagant settings, dart around shrubs in misty woodlands, sing at the tops of their voices in forest glades or on mountain peaks. Instead they met at a bus stop, office canteen, or railway station—an electrifying shift. Within this naturalistic frame Chatterji worked with the economy of the cartoonist and an eye for the small telling detail.

The Indian audience had more reason than most audiences to be obsessed with boy-girl relations. Only within the memory of people still living had the young begun to take a hand in shaping these relations. Countless marriages continued to be negotiated, and newspapers were filled with matrimonial ads placed by parents. The

[5] Interview, Chatterji.

Chatterji films really represented a romantic avant-garde, depicted discreetly but vividly. In the Indian environment they constituted a how-to genre.

The Chatterji films sometimes dissociated themselves—or seemed to—from Hindi film ritual. In several films young people go to movies whose standard style contrasts sharply with the story we are following. In *Choti Si Baat* (Just a Trifle), 1975—featuring the engaging Amol Palekar, the central figure of a number of Chatterji films—the young man tries throughout much of the film to strike up an acquaintance with a girl who rides to work on the same bus. He tries to gain courage by imagining himself a movie hero, confronting her in a misty dell, and bursting into song. He fantasizes several such ploys. Here Chatterji is slyly exploiting the standard genre while seeming to ridicule it. The young man finally manages to talk to the girl on a bus—with her help.

In several Chatterji films—*Rajnigandha* (Tuberoses), 1974; *Chitchor* (The Love Thief) 1976—the girl resolves the issue by her decision. Sometimes it is a matter of her choosing between two plausible young men, rather than of her recognizing the man of destiny. Here again Chatterji is quietly subverting tradition.

Two FFC first films made their debut in 1973. One was *27 Down,* directed by Avtar Kaul, and showing a highly sophisticated feeling for the medium. Spare of dialogue, resourceful in the use of background sound and music, it told a love story interwoven with a symphony of railroad lore and mystique. Tragically, Kaul died in a drowning accident on the day his film won the Best Hindi Feature award. It later won a Locarno festival award.

The year's other FFC-financed first feature was M. S. Sathyu's *Garm Hawa* (Scorching Wind), whose production was itself a saga. Based on a story by Ismat Chugtai, it dealt with the experiences of a Muslim family of Agra in the violent aftermath of the India-Pakistan partition—the sort of subject long considered out of bounds for Indian films. While Sathyu was filming it on location in Agra, with occasional harassment from bystanders—whom he finally managed to divert with a fake "second unit" using an unloaded camera—admiring friends advised him in a spirit of compassion that it could never be exhibited.

BASU CHATTERJI's *Rajnigandha* (TUBEROSES), WITH VIDYA SINHA AND AMOL PALEKAR, 1974

BASU CHATTERJI's *Swami* (LORD AND MASTER), WITH SHABANA AZMI AND UTPAL DUTT, 1977

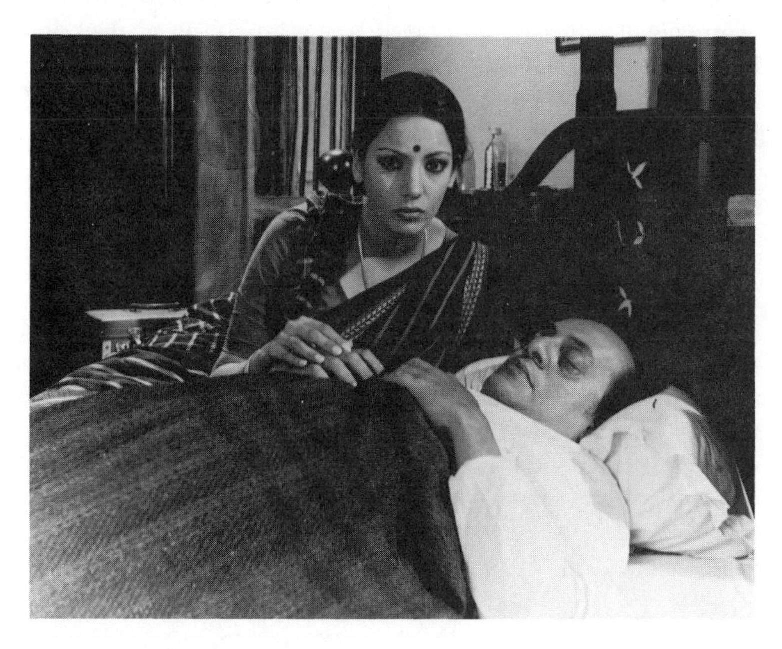

Sathyu, a Hindu from Karnataka,[6] had a background in the graphic arts, which led to his involvement in regional theatre as a scene and costume designer. Moving to Bombay in 1952, he acquired a job designing advertising cartoon films for Lifebuoy soap and other products. He studied film via short-term jobs in various productions—as costume designer, general assistant, production manager. In this way he acquired a firm grasp of production techniques before looking for backing for a feature film. His plan for *Garm Hawa* was at first accepted by a commercial producer, who then became fearful and backed out. Application to the FFC followed. The veteran actor Balraj Sahni agreed to play the central role—the last before his death.

The role was that of Mirza Salim, a shoe-manufacturing Muslim of Agra who is confident that the occasional signs of hostility following the Partition will "pass off in a few days." But his optimism is short-lived. Some Muslims have fled to Pakistan without notice, so banks decline further loans to Muslims, and buyers hesitate to place orders for the same reason. As Mirza's business declines, his employees start to leave, and the economic crisis leads to other crises. Some family members leave for Pakistan, but Mirza is determined to stay. Short episodes, directed with precise control, trace the gradual disintegration of his hopes. Finally he and his son also head for the border. But on the way the son is deflected by a protest parade led by red flags—and joins the parade.

True to prediction, the censor board at once banned the film as an instigation to communal dissension. But Sathyu had taken care to show the film to numerous government leaders and journalists, and their indignation over the board's censorship finally led to a reversal of the ban. The film eventually won a National Award for its contribution to "national integration." Thus another career was launched via FFC financing.

Sathyu's second film, *Kanneshwara Rama* (The Legendary Outlaw), 1977, again showed his penchant for carefully architectured work grounded in social issues. It was set in the early twenties, when Mahatma Gandhi's non-violent movement was gaining momentum.

[6] The Kannada-speaking area, formerly Mysore State.

M. S. SATHYU'S *Garm Hawa* (SCORCHING WIND), 1973: LAST ROLE FOR BALRAJ SAHNI

The film focused on the rival attractions of Gandhism, regarded by many as an impossibly idealistic challenge to empire, and terrorism, as represented by a folk hero of Sathyu's home area, Kanneshwara Rama—a subject of folk ballads still sung in Karnataka. The folk ballads provide the drama with a narrative framework. Made in the Kannada language, the film had a potential audience far smaller than that of the Hindi *Garm Hawa,* Sathyu's first feature.

FFC-financed first films continued to appear, ranging widely in skill and perception and in industry acceptance. Some remained undistributed. Meanwhile several states were launching film-support plans that had a similar impact, although the purpose was different. In 1965 Karnataka, wishing to foster film production within the state, launched a unique subsidy plan. A feature film shot mainly in Karnataka would, on securing a censorship certificate, automatically receive a cash subsidy. No script review was involved. Subsidies were at first set at Rs. 30,000 per Kannada feature but by the 1970s stood at Rs. 100,000 for a black-and-white film, and Rs. 150,000 for a color film. Smaller subsidies were offered for films shot within the state in other languages. In addition, a film found to have unusual cultural value could be exempted for a time from the state entertainment tax, so that theatres could show it at sharply reduced prices. Furthermore, outstanding work would be rewarded with cash prizes.

The incentives caused a new film industry to rise, making Bangalore rather than Madras the center of Kannada-language production. The new enterprise drew the interest of leading Kannada-language writers and theatre talent, which gave the films a strong regional quality of the sort that had marked the Bengali films of Satyajit Ray and had made them universal.

In 1970 *Samskara* (Last Rites), a film by Pattabhi Rama Reddy about Brahmin rituals and prejudices of earlier decades, became an important influence. Because of its anti-Brahmin implications it encountered a censorship ban, but again a surge of indignation brought a reversal. It won wide admiration for its ethnographic authenticity, and set the tone for other such works.

An important follower of Reddy was B. V. Karanth. Having begun his career as a theatre errand boy, he had become a leading director of the Kannada theatre as well as a music scholar and composer. At this point he was virtually thrust into film production. A writer agreed to sell his story for a film version only if Karanth would direct it. The result was the 1971 release *Vamsha Vriksha* (The Family Tree), which Karanth persuaded the dramatist Girish Karnad to help him direct. The experience confirmed Karanth's interest in film, and he headed for the Film Institute at Poona to study the medium and look at archive classics. By 1975 he felt ready for his first solo production, *Chomana Dudi* (Choma's Drum), a remarkable chronicle of a Harijan family of a half-century ago. Choma, the father, has a yearning for a small parcel of land and a latent fury he can articulate only through his drum. We see his hopes disintegrate as he loses one son to the river, another to cholera, a third to Christianity. A nearby Christian mission is offering Harijans small parcels of land if they become Christians, a condition which seems to many Harijans to offer a lower status than their own.

While the son succumbs to the offer, Choma does not. At the end, defeated, he dies. But his angry drum is still heard on the sound track.

Karanth became director of the National School of Drama in New Delhi, with theatre as his main arena, but he retained an interest in film, and sometimes worked as a film composer. He typified the strong link between the rising Kannada film movement and the regional theatre.

Girish Karnad was already a well-known playwright when he had his first contact with film as the leading actor in *Samskara*. He then wrote the screenplay for *Vamsha Vriksha* and collaborated with Karanth on its direction. He too was now irresistibly drawn to film and in 1973 released his first solo production, *Kaadu* (The Forest), which was brilliant in its conception and execution. With precise control over intimate scenes as well as crowd scenes, he showed exceptional skill in the handling of children.

Kaadu depicts two villages separated by a forest and a smoldering feud. The chieftain of one village is involved in a liaison with a woman in the other village, and goes to her at night. His wife, visiting a magic practitioner, performs a ritual to win him back. Hostility between the villages mounts, and a traditional festival is organized to bring about a reconciliation, but instead it leads to a violent climax. An absorbing aspect of *Kaadu* is that these conflicts are seen through the eyes of a child, the chieftain's son, giving them an ironic impact.

The authority of Karnad's work won him leading roles in a number of films by other directors, and also a stint as director of the Film and Television Institute of India. By 1978 he was ready for another major film of his own, *Ondanondu Kaladalli* (Once upon a Time), released early in 1979. The film confirmed his status as a director of the first rank. It dealt with the Karnataka area in early medieval times, when chieftains vied for domination over sparsely populated forest areas. The men of the region were famous for their skills in the martial arts, and its greatest exponents were sought far and wide as mercenaries. An army consisted of a few dozen such warriors, and power could shift with the acquisition of a leading champion of the moment, resulting in new crises and tribal realign-

GIRISH KARNAD'S *Ondanondu Kaladalli* (ONCE UPON A TIME), 1979

ments. The film, seductive in the beauty of its forest landscapes, was also spectacular in its martial action. Karnad made his story startlingly believable in its own primeval terms, yet it had overtones of the arms races and shifting power alliances of other eras.

The success of Karnataka film policies led to similar incentives in other states, notably in Kerala. The most densely populated state, and also the one with the highest literacy rate, Kerala was often in political ferment, a condition which had influenced its film making since the mid-1960s. At the same time it experienced an economic boom, resulting from remittances from Kerala workers employed in the oil countries of the Persian Gulf. Some of the boom profits went into film production, and a state film-subsidy plan, though modest in size, added fuel to this. The overall result was a Kerala film-making explosion, mostly in the Malayalam language, from which a number of promising film talents began to emerge.

Among them was G. Aravindan. Like Basu Chatterji and M. S.

Sathyu, he had a background in the cartoon field, and showed a similar feeling for precise detail and expressive vistas. He too was strongly influenced by film-society stimuli, as well as by the politically charged atmosphere of Kerala—the first Indian state to elect a communist government. In Aravindan's first feature, *Uttarayanam* (Solstice), 1974, a young man reflects on the past struggles of the anti-British freedom fighters he has learned about from his paralyzed father. He wonders what obligations their sacrifices place on him today, while he watches new establishments replace the old, bringing similar inequities. In his second film, *Kanchana Sita* (Golden Sita), 1977—again in Malayalam—he had a story of the *Ramayana* re-enacted by a tribal group who regularly visited Trivandrum to sell herbal medicines. The film gave the familiar story, usually depicted as a magnificent, princely costume drama, a startlingly new context. With photography, editing, and sound recording by Film and Television Institute alumni, the film was another young-talent breakthrough.

Another product of the Malayalam film explosion was M. T. Vasudevan Nair, a distinguished journalist who made his film debut with *Nirmalyam* (Flower Offering), which won the 1973 Best Feature award. A village story of superstition culminating in an extraordinary dramatic climax, it exhibited a firm directorial hand and a dedication to cultural authenticity.

But the increase of films in the Kannada and Malayalam languages raised problems. In 1978 films in Malayalam actually outnumbered films in Hindi 123–122, exceeding the capacity of the market to absorb them. Films in Kannada and Malayalam faced the same problems that those in Bengali had to contend with: a film might pay its way in the home area, but was likely to reach other language areas mainly through special "morning" shows, which yielded little revenue. It might get more attention abroad than in India as a whole. This problem kept luring directors into production in Hindi, the one language that, in spite of regional antipathies, had a quasi-national reach. This gave particular importance to a film maker who emerged in 1974 with a spectacular first film in Hindi—*Ankur* (The Seedling)—which at once gave him national prominence.

Shyam Benegal came from the Hyderabad area, in Telugu-speaking Andhra Pradesh, which was to serve as the locale for his first feature. He eventually settled in Bombay. Events leading to his film career had striking similarities to those involving Satyajit Ray. Like Ray he had studied economics and helped found a film society—the Hyderabad Film Society—later attending another, the Anandam of Bombay. Like Ray he had an older cousin who was a widely respected film director—in this case Guru Dutt—and Benegal sometimes accompanied him to the studios. Like Ray he had made a valued sojourn abroad—in Benegal's case, working at a television station in Boston and the Children's Television Workshop in New York. Like Ray he went into advertising. Like Ray he began to think about making a feature film, drafted a scenario, and tried for years to raise funds. No backer seemed interested.

Over the years Benegal's work involved the making of innumerable commercials and occasional sponsored documentaries. He had a high standing in the commercials field. Then Blaze Advertising, which dominated the booking of commercials into Indian theatres and had surplus funds to invest, made a surprise suggestion. Planning to enter the feature field as Blaze Enterprises, they proposed backing a Benegal feature production. Benegal retrieved his scenario from the closet and went to work. The result was *Ankur,* which propelled into fame not only Benegal but also the young actress Shabana Azmi, a Film Institute alumna.

The film reflected the rigorous technical training his commercials had given him, but also careful thought and acute perception. Based on an actual episode in the Hyderabad area, the film was made with the ethnographic fidelity prized by many of the new directors. His direction showed a sure, subtle control. Women were directed especially sensitively and his handling of crowds suggested a seasoned virtuosity.

Ankur, like many of Satyajit Ray's films, focused on power and privilege, and avoided doing so in hero-villain terms. We meet the attractive son of a landlord and sympathize with him when he is sent to administer a rural estate to put it on a paying basis, when he would rather be elsewhere. He is to live at the estate until his young wife by an arranged marriage is ready to join him. He reluctantly accepts the situation and tries to understand the workings of the estate and the erosion of its profits. It seems not unnatural that he is attracted to a pretty servant girl on the estate, and she to him. A liaison develops—a standard event in zamindari lore. He treats her well, letting her move from her hut to the main mansion. She stays there until the wife arrives to join him. Then the servant girl returns to her hut.

Her husband is a deaf-mute who has been away many months, having left in disgrace over a pilfering incident. When he returns, rejoining her in the hut, she reveals that she has become pregnant. To her surprise he rejoices, and seems to contemplate a family life. When the deaf-mute goes to pay his respects at the mansion, and to ask for re-employment, the young landlord mistakes his purpose. As the deaf-mute approaches, the young man grows fearful. Anticipating exposure, and perhaps an attack, he lashes out at the deaf-mute and beats him savagely as other workers stand by in horror without lifting a finger. At the end of the film a small boy throws a stone at the landlord's window.

That final symbolic act of rebellion foreshadowed much in Benegal's work. Later films would have rebellious acts of a more decisive sort. Acceptance of fate is not a Benegal theme. In his work, and its appeal to the young, we sense that doomed heroes and heroines, so long a fixture in Indian films, are doomed.

Benegal's second feature, *Nishant* (Dawn), 1975—impressive but less successful than his first—concerned a zamindar of the pre-independence period, which is referred to in an opening title as a "feudal" time.[7] He is more a traditional villain, but the focus in this film is on the corruption that comes over lesser leaders of the village—priest and police officials—as they accommodate their decisions to his power and produce grotesque injustices. In this case the village ultimately rebels and in a riotous sequence overwhelms and slaughters the zamindar family.

Benegal, like Ray, tried diverse techniques and genres. In 1975 he made a children's film, *Charandas Chor* (Charandas the Thief), and then undertook a larger project unique in Indian film annals. The leader of a Gujarat milk cooperative urged him to make a film

[7] The zamindari system maintained under the British was abolished by Indian legislation soon after independence.

SHYAM BENEGAL'S *Ankur* (THE SEEDLING), 1974,
WITH SHABANA AZMI AND ANANT NAG. BLAZE

about the rural cooperative movement, and offered to secure backing. Eventually 500,000 members of the milk cooperative invested two rupees each, providing a budget of Rs. 1,000,000. The film became *Manthan* (The Churning), in which the opening title read: "500,000 farmers of Gujarat present . . ." When it was released in 1976, the investing farmers came to Bombay in huge busloads to see "their film" and helped to turn it into a commercial success.

Girish Karnad appeared in *Manthan* as an organizer attempting to form a new cooperative. After initial successes the cooperative faces destruction resulting from organizing errors, commercial opposition, and internal caste conflicts. He leaves, accepting defeat, but some of those he leaves behind are determined to pursue the project. The plot carefully avoids overemphasis on a hero's leadership.

Bhumika (The Role), 1977, involved still another change of pace. A fictionalized reenactment of the life of a celebrated, rebellious screen actress of an earlier decade, it offered fascinating glimpses of films and studio life of the period, and of the status of women. Its structure was diffuse, but its impact powerful. Its lead role was played by Smita Patil, a former news reader for Indian television whom Benegal had cast in film roles in *Nishant* and *Manthan* and who became a top-ranking star on the basis of her *Bhumika* performance.

Benegal made his fifth feature, *Kondura* (The Sage from the Sea), 1977, in both Hindi and Telugu, the language of his home area. In this film he returned to the rural scene, in a drama of dogma and superstition somewhat reminiscent of *Samskara*. Meanwhile he was approached by Shashi Kapoor, who had become a superstar of the Bombay film world, with an offer to finance a feature in which Benegal would have total control. Benegal accepted, received his largest budget to date—rumored at Rs. 6,000,000—and decided on a drama of the 1857 Sepoy rebellion, for which he assembled a stellar cast including Shashi Kapoor, Jennifer (Kendall) Kapoor, and Shabana Azmi. The trade press asked: Is Benegal "selling out" to the establishment? They could have asked: Are the new film makers taking over the establishment?

Junoon, based on a story by Ruskin Bond titled *A Flight of Pigeons,* emerged as a historic epic with vivid action and an absorbing human story—an Indian *Gone With the Wind.* Not surprisingly, it was selected to open the 1979 International Film Festival of India, and went on to successful commercial runs.

The excitement that Satyajit Ray had created in 1956 had spread far and wide. The rise of Benegal had its symbolic meaning. Dissent had reached firm commercial footing.

Some called it a "movement," but that seemed a misnomer. The new film makers had pushed in many directions, united mainly in their aversion to formula. Rejection of industry dogma was their main dogma. Some had esoteric leanings, but most followed Ray in disavowing the "minority audience syndrome." They wanted to express something to a wide audience. The economics and potential reach of cinema seemed to them to rule out any other approach.[8]

The ferment had touched the documentary field. In 1965–67 Jean Bhownagary, who had been involved in early Films Division history

[8] Ray, *Our Films, Their Films,* p. 98.

POSTER FOR *Manthan*

and had then gone on to a UNESCO position, returned as Chief Adviser to the Films Division and generated an adventurous spirit, particularly in planning material for the twentieth anniversary of Indian independence. For this occasion the Films Division caught the *cinéma vérité* fever that was sweeping all continents, which resulted in such Films Division projects as S. N. S. Sastry's *I Am Twenty,* a beguiling collage of impromptu statements by twenty-year-olds on their hopes and fears for India; and S. Sukhdev's *India '67,* a witty, stimulating, sardonic—and at the same time inspiring—juxtaposition of footage, music, and sounds from all parts of India, which identified achievements as well as continuing problems. Unfortunately, it reached theatre audiences only in a truncated twenty-minute "approved film" version entitled *An Indian Day.* Other outstanding documentaries of the period included Fali Billimoria's impressive *The House That Ananda Built* and P. D. Pendharkar's *Weave Me Some Flowers,* on Indian textiles—a Films Division excursion into sophisticated design that was free of condescending narration.

The *cinéma vérité* technique later infiltrated regional cinema in several projects, including the semi-documentary Malayalam feature *Thampu* (Circus Tent), 1978, by G. Aravindan, intertwining fictional passages with highly evocative documentary footage of a traveling circus and its audiences as it wound its way through Kerala villages. Another was the Merchant-Ivory *Autobiography of a Princess,* 1975, featuring Madhur Jaffrey and James Mason in a dramatized framework with flashbacks of maharajah home movies. Still another unusual project was *Where Centuries Coexist,*[9] two feature-length films by S. Krishnaswamy surveying Indian history through current sights and sounds, filmed and recorded throughout India. Originally projected for foreign television, its Indian distribution rights were purchased by Warner Brothers in 1976, winning it coverage in Indian theatres usually devoted to foreign films.

The independently produced *The Framework of Famine,* made in 1967 by Pratap Sharma, not only presented a stark picture of famine in Bihar but also examined causes of recurring famines in the area in a manner that indicated that corruption was at least partly to blame. In spite of official fears and obstructions, the film finally received high-level approval, and went into Films Division distribution.

The period of the late 1960s and 1970s had been one of probing, with innovation moving from the fringes toward the center of the industry. Satyajit Ray had been joined by such rising talents as Shyam Benegal, Mrinal Sen, Girish Karnad, M. S. Sathyu, Basu Chatterji, B. V. Karanth, M. T. Vasudevan Nair, Pattabhi Rama Reddy, G. Aravindan, and others.[10] They had experimented with form and technique and widened the subject matter of Indian films. Some were political to an extent that had not seemed feasible in earlier years.

It had been a period of new birth. In the history of the arts, such

[9] Originally *Indus Valley to Indira Gandhi.*

[10] Others mentioned in the 1970s: in Gujarati films, Kantilal Rathod; in Hindi films, Mani Kaul, Saeed Mirza, Hrishikesh Mukherjee, Shivendra Sinha; in Kannada films, Girish Kasaravalli; in Malayalam films, Adoor Gopalakrishnan, Ramu Kariat; in Tamil films, K. Balachandar, Jayakanthan; in Telugu films, Bapu, Singitam Srinivasa Rao.

periods have generally been brief, vulnerable to changes in political or economic climate. There was a special vulnerability in the new Indian upsurge, in that it had received important stimulus from innovations in government policy. What would be the effect of shifts in policy? This question was forcefully posed by a brief and bizarre chapter in Indian history, the Emergency.

Emergency

It was a time when leaders who opposed government policies—including such elder dignitaries as Jaya Prakash Narayan and Morarji Desai—were suddenly awakened in the middle of the night and taken to prison. It was June, 1975. In the roundup that followed, tens of thousands of their followers, including members of Parliament and state legislatures, were also jailed, and some were tortured. Under Emergency edicts and hastily enacted laws, reasons for arrest became "confidential matters of state" which must "not be communicated to the person detained." Appeal through the courts was outlawed. Information about arrests was not to be published. A former Deputy Minister of Defence, V. C. Shukla, installed as Minister of Information and Broadcasting to replace the liberal and popular I. K. Gujral, took firm control of the news media, ordering what must and what must not be reported. News of price rises, the bulldozing of hutments, or hunger strikes became "objectionable matter," and publication about these topics could mean arrest for journalists. Troublesome newspapers were forced to place government nominees on their boards, replacing others. Some periodicals, such as the notable *Seminar,* decided to cease publication. The news media became a propaganda engine extolling Emergency decrees and achievements, and news acquired a rosy hue. At first generating acceptance, it soon produced a vast rumble of underground rumor, fed by clandestine broadsheets. Government efforts to suppress these became unrelenting. Seven thousand people were arrested for printing and circulating broadsheets. On the surface the rosy hue prevailed, producing deception along with massive governmental self-deception. The ministries read what they churned out.[1]

[1] For a detailed treatment see Mehta, *The New India.*

The proclaimed Emergency lasted almost twenty-one months. It had devastating impact on the press. It had less effect on All India Radio and the Films Division because long habit had made them obedient servants of the ministries. Controversial issues were regularly handled by quoting current ministers. The Films Division had generally acted as though it were a public relations firm with one client: the current administration. Since the Congress party had held power so long, this had been an easy habit to acquire. When the Films Division or the broadcast media were instructed to glorify the Emergency, to issue extravagant claims and ignore the objectionable, it was merely a continuation of already standard procedure. In their eagerness they tended to exceed and anticipate instructions, as in their constant featuring and quoting of Sanjay Gandhi as presumed leader of the nation's youth, and perhaps heir-apparent.[2] And they needed no instructions to quote Prime Minister Indira Gandhi constantly, particularly on discipline: "Discipline makes a nation great."[3]

For the Films Division it was business as usual, but with an extra urgency. The Prime Minister proclaimed a Twenty-Point Program rationalizing the Emergency decree and defining its aims. Twenty film makers were summoned to a Bombay meeting with officials of the Ministry of Information and Broadcasting and instructed to make a series of "Twenty-Point Films"—one film to a point—for release to theatres under the "approved film" program. Each film maker was assigned one point. Fourteen films were completed and distributed. Their main titles carried a standardized design featuring a large 20. Some dealt with programs already in effect before the Emergency, such as plans for new housing, and seemed intended to give the Emergency credit for them. Other films made absurdly tall claims: the Emergency, said one film, had at last ended—or virtually ended—the practice of selling girls into prostitution.[4]

When the Emergency ended with the victory of the Janata party and the ousting of Prime Minister Indira Gandhi, the Films Divi-

[2] For a Films Division example see *Dialogue: Youth Power,* a 1976 "approved film."

[3] *The Question,* a 1976 "approved film."

[4] *After the Silence,* a 1976 film in the Twenty-Point series.

sion withdrew the Twenty-Point films. There was perhaps no reason to scuttle some of them, such as those attacking black market operations and smuggling, except that they were associated with the defeated regime and its Twenty Points. The Division had a new client.

At the end of 1977—year of the Indira Gandhi defeat, the liquidation of her Emergency, and the rise of Janata—the Films Division produced a documentary summarizing the year's events, and managed not to mention Indira Gandhi. The *Hindustan Times* reported that the Janata party was "considerably embarrassed by this exhibition of zealous loyalty," and denied having inspired it. The documentary was ordered withdrawn.[5]

The feature film industry felt the hand of the Emergency occasionally, mainly in the form of unusual censorship edicts. Several of these reflected the paranoia of the Ministry of Information and Broadcasting.

Mohan Bhatiya, a manufacturer of paper boxes and other products, financed and produced the Hindi feature *Andolan* (Revolution), completed early in 1975, before the Emergency. It concerned freedom fighters of the 1942 period, the time of the Quit India movement. The film was previewed at Bhatiya's invitation by several government leaders, and won an accolade from Jagjivan Ram, a longtime cabinet minister,[6] who praised it for its educational value: "It vividly portrays the massive response of the people to the clarion call of Mahatma Gandhi." In May, 1975—a month before the Emergency proclamation—the film won censor approval with an "educational" rating, which usually resulted in a waiver of state entertainment taxes. Such a waiver was recommended by the Ministry of Information and Broadcasting, at that time headed by I. K. Gujral. A premiere was scheduled for mid-August. Meanwhile the Emergency was proclaimed.

A week before the scheduled premiere, Mohan Bhatiya received a phone call from the Ministry of Information and Broadcasting, by this time headed by Shukla, ordering him not to release the film,

[5] *Hindustan Times*, February 28, 1978.
[6] In the Janata regime he became Deputy Prime Minister and Minister of Defence.

but to despatch a print to the Ministry for study. The censorship certificate was not withdrawn, but exhibition was forbidden, with hints of reprisal if orders were disobeyed. Bhatiya, whose paper box business depended on government patronage and credit from the nationalized banking system, dared not ignore or challenge the ban. The Ministry took no further action, and the film gathered dust. When, after the Emergency, it finally opened in uncut form, it quickly failed. What had so terrified the Shukla regime?

Andolan was fully in the Hindi film pattern. With songs and dances, it glorified rebels fighting oppressive British overlords. The hero sings while staggering with a gun wound, and while being whipped, marched to jail, in jail (backed by a chorus of other prisoners), as well as on the way to the gallows, and at the gallows. During his final playback song the noose frames his head. As he sings, it is placed over his head. As his body plunges, the defiant song is jerked to a halt. Then we see his flower-decked body carried away, with a reprise of the song on the sound track. The film is about a rebellion in never-never land, in which villains torture and kill, while heroes suffer, raid a police station, and derail a military train. Apparently the Shukla ministry feared the film would incite resistance to Emergency rule. That the ministry should thus identify itself with tyrants in a musical fantasy was perhaps more bizarre than anything in the film.

A somewhat similar case was *Kissa Kursi Ka* (The Story of a Chair), a fiction drama about government corruption and the misuse of power, produced by Amrit Nahata, a member of Parliament, also completed before the Emergency. Because of a long delay in the granting of a censor certificate, Nahata went to court. In August, 1975, the Ministry seized the negative and all prints and banned the film *in toto*. Subsequently the Supreme Court, reviewing the case, asked the Ministry to bring the film to court. The Ministry apparently decided to avoid this by destroying the film, and reporting to the Supreme Court that no print could be located. According to later testimony, the film was incinerated in a pit of the Maruti automobile factory, the enterprise launched by Sanjay Gandhi.[7]

[7] *Times of India,* January 12, 17, 1979.

The film was eventually reshot. Like *Andolan,* it created no sensation and quickly failed. Again, the film apparently offered nothing as remarkable as the Ministry's reaction to it.

In Karnataka the makers of *Samskara* felt the hand of the Emergency in a more cruel fashion. Snehalata Reddy, the leading actress in *Samskara* and wife of its director Pattabhi Rama Reddy, was accused of concealing information about the whereabouts of George Fernandes, a trade union leader whose arrest had been ordered in the Emergency roundup.[8] She was known to be a friend of the Fernandeses. She denied knowledge of his whereabouts and was jailed and questioned for eight months. An asthmatic deprived of needed medicines, she fell seriously ill, and was released only when near death. She died in January, 1977—five days after her release.

In 1977 the Prime Minister, confident of a sweeping victory, called for "fresh elections" and reaped disaster. The nightmare ended as suddenly as it had begun. The numerous laws enacted to implement the Emergency were repealed under the Janata regime. The film industry was where it had been, but perhaps with a more acute sense of its problems.

This showing sold out

The Indian film world is conscious of pressures for change, but they continue to be outbalanced by forces for the status quo. The main reason can be simply stated: theatres are crowded. Throughout India lines of people wait at box offices. No other medium matches film in its hold on wide audiences.

Since the earliest film days these audiences have included a broad range of economic strata, which are reflected in the range of admission prices. Three, four, or five price brackets are common. In the average city cinema, seats for evening showings may range from a low of Re. 1 to a high of Rs. 5; in the fancier city cinemas the highest-priced seats may run considerably higher. In mofussil, the rural areas, the range is likely to be lower. In some rural theatres only the highest-priced seats are chairs, while some of the lower-

[8] He became Minister of Industries in the Janata government.

priced seats are benches with backs, and still lower ones are benches without backs. In earlier years some people also sat on mats on the floor, but this practice has practically vanished.

The Westerner accustomed to visiting a cinema on the spur of the moment may find this difficult in India. In the cities the better seats for weekend showings are often sold out days ahead. Evening showings of a successful film may also be largely sold out ahead of time, and tickets may have become a black-market item. All except the cheapest seats are sold on a reserved-seat basis. One buys a particular seat for a particular showing. In the big metropolitan cinemas there may be four showings, at approximately noon, 3, 6, and 9 P.M. On Sundays and even on Saturdays a fifth showing may be added, at about 9 A.M. At cinemas in less crowded areas there may be only two showings per day, in most cases after sundown.

In big cities the 9 A.M. and noon showings may be devoted to films other than the scheduled feature. These may be revivals, children's films, foreign films, or films from other language areas, including films by the new film makers, sometimes referred to as "parallel cinema." The films of Satyajit Ray first appeared in non-Bengali areas in "Sunday morning shows"—a scheduling practice that later spread to other mornings. The 9 A.M. and noon showings are often called "matinees." Newspapers may carry special lists of "matinee" films.

Since a theatre must be emptied, then filled again, in the brief gaps between showings, there is likely to be much congestion at this time. A half hour before the break the lobby begins to be crowded with waiting people. The city crowd has a varied look: some wear the traditional dhoti, some are in pajamas, some in pants with an Indian shirt or closed coat, and some wear Western clothes. Some are barefoot, some wear sandals, some shoes. Most women wear saris; ladies in elegant silk saris with much jewelry are in evidence, but many younger women wear blue jeans. There are usually many family groups.

Among waiting people some squat, suspended an inch from the floor. Some stand about. Many theatres have food counters or restaurants, where the more well-to-do may have a snack while waiting, or during the intermission. Outside there is a coming and going of taxis, tongas, three-wheeled motor rickshas, and bicycle rickshas. The lobby features stills of current and coming attractions, and perhaps a large cutout display of a star. The emphasis on the star is everywhere in evidence. Fan magazines—there appear to be some seven hundred in India—and song leaflets may be hawked on the sidewalk and in the lobby.

While city cinemas are architecturally similar to Western cinemas, some rural cinemas are different. In the south they may be thatched-roof structures, which offer several advantages. An excellent insulator, thatch offers cool shelter. It also has fine acoustic qualities. Generally operating after dark, and open at the sides, it offers little fire hazard. The projection booth, at the rear, has its own tin structure. The cinema is surrounded at a slight distance by a fence or stockade; admission is paid when entering the enclosed area, which includes a refreshment stand as well as latrines.

Thatch loses its insulating effect in due time, so the cinema may after two or three years be torn down and rebuilt in the same spot, perhaps after the monsoon. In effect it is an old-style traveling cinema that has settled down, and may still operate under rules for "temporary cinemas." These give escape from some of the standards imposed on city cinemas. States have allowed this because cinemas showing Films Division films and sometimes state documentaries are a government communication channel. Operating at low economic level, many could not survive except as "temporary" cinemas.

Cinemas also serve as a business communication channel, via filmed commercials and slides. Sale of screen time for advertising has burgeoned in recent years. In rural cinemas, commercials for fertilizers became prominent in the 1970s, and are said to have played a part in the "green revolution." In city cinemas the commercials resemble television commercials, promoting soaps, cosmetics, clothing, appliances, motor scooters, television sets, and other consumer items—many in the semi-luxury class. Their style often contrasts sharply with that of the audience. They appear to be a Westernizing pressure, which some fear may widen the gap between city and village in economics, culture, and outlook. The entertainment has a parallel merchandising impact, especially through the clothes, hairstyles, and mannerisms of stars and their lifestyle as reflected in fan magazines.

At the cinemas the commercials are grouped with the Films Division documentary or newsreel at the start of the show, and are generally accompanied by the rumble and shuffle of people finding their seats. This noise stops abruptly as the feature begins.

While stars are the main obsession of cinema devotees, there are other deities in the pantheon. The music director—usually a composer-arranger-conductor—continues to have a status second only to that of the star. His collaborator, the lyricist, generally occupies a slightly lower rung, but is still idolized. He may be listed in the credits as "poet," and may be a well-known published poet, as in the case of Gulzar, who went from poetry-writing to lyric-writing to script-writing to directing. Dancers may likewise be idolized; for some, dance provides the road to stardom.

Another idol—a remarkable Indian phenomenon—is the playback singer. In most countries his or her very existence is kept secret, with the assumption that audiences must be persuaded that the star does the singing. The Indian film world scorns this deception. Playback singers are identified and have their worshipful following. Most astounding has been Lata Mangeshkar, for more than three decades an unseen participant in hundreds of films, and invariably listed in the credits, singing thousands of songs for which others have moved their lips and tensed their throats—songs in Hindi, Punjabi, Marathi, Gujarati, Rajasthani, Bengali, Konkani. Lata Mangeshkar phonograph records of these songs, under the Indian "His Master's Voice" label, have sold in the millions. In most recording sessions her first take has been the final take, performed after only a brief perusal of the score. Her brother has long been a composer, and her sister Asha Bonsle is a playback singer, ranking second only to Lata.

The script writer has not generally enjoyed the status he holds in other countries. Many Indian directors have continued to write their own scripts, and some have preferred to do without scripts, working from a rough synopsis and improvising as they go.[1] The

[1] Mehboob Khan, the ebullient Bombay producer especially known for his *Mother India,* prided himself on not needing writers. He said he had tried writers once, seven of them simultaneously, but they sat around looking so solemn that they unnerved him, and he fired them all. Inasmuch as he had had many successes without them, he saw no reason to doubt his procedures. Interview, Mehboob Khan.

BOMBAY: MEHBOOB KHAN'S *Mother India,* STARRING NARGIS, 1957. MEHBOOB PRODUCTIONS

importance of the script has grown in the "parallel" cinema, but here too it is often the work of a writer-director.

The host of others receiving screen credits may include such arcane specialists as the "director of fights." G. D. Khosla, who chaired a committee reviewing the workings of censorship, commented: "The inclusion of a 'director of fights' in the credits of a film is a recent and deplorable innovation which invites the viewer to regard fights as some kind of art and aesthetic entertainment."[2]

But all such specialists are overshadowed by the music specialists, whose prestige is related to the key role of music in the apparatus of

[2] Khosla, *Pornography and Censorship in India,* p. 160.

fandom. Film songs continue to be released for radio use before the release of the film itself. If all goes well, songs are hits before the premiere, bringing murmurs of excited recognition as the film unfolds.

The status of film songs has changed since the days when All India Radio condemned them as a corruption of the Indian ear. Because the ban made Radio Ceylon, with its schedule of film songs, the favorite throughout India, AIR was forced in the 1960s to institute Vividh Bharati, its "light music program" featuring film songs. The advent of the transistor radio contributed to the success of Vividh Bharati, which quickly overtook Radio Ceylon. The radio audience grew substantially during the 1960s and 1970s. All this further extended the sovereignty of film songs, which have gradually pervaded all aspects of Indian life—accompanying weddings, funerals, state occasions, religious festivals, parades, parties, and political conventions. As television began in major cities, the most popular series became compilations of song-and-dance sequences from old and new films—a formula adapted with success to various language areas. Among the cognoscenti, denouncing film songs is no longer as fashionable as it once was, and musicologists have taken increasing interest.

These comments of Satyajit Ray are significant:

> I have been able to watch the development of the Hindi film song over the years thanks to my son's continued interest in them. I keep being amazed at the inventiveness that is poured into them. As poetry they are often no great shakes but how many songs ever are? In fact one of the conditions of a good lyric is that it should not be great poetry because a great poem carries its own charge of music. But the really striking things are in the tunes and in the orchestrations. They embrace all possible musical idioms—classical, folk, Negro, Greek, Punjabi, Cha-Cha, or anything you can think of from any part of the world. The latter shows a brashness and a verve in the combination of instruments—again as disparate as you can imagine—and a feeling for tonal colour and contrast which call for high praise.[3]

The role of music reminds us that Indian cinema is not really comparable to cinema elsewhere. With television still a very minor

[3] Ray, *Our Films, Their Films,* pp. 74–75.

factor, and other sources of entertainment not available to most of India's millions, the cinema has multiple functions. In many regions folk festivals and performances have begun to yield to the advance of cinema, and have to some extent been replaced by cinema. Describing Indian cinema at a Venice film festival, Sehdev Kumar Gupta called it "at once a nightclub and a temple, a circus and a concert, a piazza and poetical-symposium."[4] It is also a music hall, a community center, an arena where good and evil clash. And all of it involves stars—a phenomenon with endless ramifications.

A strange aspect of Indian stars is that devotees, fan magazines, and even critics generally refer to them as "film heroes" and "film heroines" rather than as "actors." An actor having achieved "film hero" or "film heroine" status is usually afraid to accept roles that blur the image. Actors thus tend to become stereotypes, and the scripts are built around the stereotypes. Public persona fuses increasingly with screen image.

A fascinating example of this phenomenon was the transformation of M. G. Ramachandran from "film hero" to Chief Minister of Tamilnadu, with dramatic crises en route. Film roles as folk warrior against evil usurpers apparently infused his political image as southern champion against the "oppressors" of the North—the establishment in New Delhi. His political tenets were vague. In later years he proclaimed Annaism as his central doctrine—in memory of C. N. Annadhurai, the early leader of the Dravidian movement, who had died of cancer. Ramachandran defined Annaism as "the best of capitalism combined with the best of communism."

Ramachandran films invariably had villains, often played by the actor M. R. Radha. Shortly before the 1967 Tamilnadu state elections M. R. Radha visited Ramachandran, who was then a candidate for the state legislature. Two shots were fired and Ramachandran fell with a bullet in his neck. Radha, who also had a bullet wound, was arrested. Each claimed the other had shot first, but no one—least of all the jury—believed the villain, and he was sentenced to jail. The hero, as in countless movies, survived the battle as his fans knew he would. Campaign posters showed him with his head in bandages, and he was triumphantly elected. Fact and fancy had

[4] Gupta, *Address,* p. 3.

once more reinforced each other, and continued to do so. Radha emerged from jail with the announcement that he would return to production with a film to be titled *Nanthan Sutten!* (I Shot Him!) As for Ramachandran, he continued to perform his hero role, both in politics and film, overcoming villains against great odds.

What is one to expect of a drama tradition so closely meshed with a hero-villain pantheon? As Professor Satish Bahadur of the Film and Television Institute put it, dramaturgy and stereotype are "firmly interlocked in the stable equilibrium of a vicious circle."[5] The new film makers who were attacking this equilibrium were fighting more than an aesthetic battle.

The social and psychological ramifications of film formulas are not easily defined. The psychiatrist Harvey R. Greenberg, in his book *The Movies on Your Mind,* writes:

> I regularly use cinema as a Rorschach test. During the first few sessions I ask about favorite films, and invariably some heretofore well-concealed aspect of psychic geography will be thrown into bold relief.

He explains:

> The movies, like waking dreams, interpret every aspect of our lives—the unquiet past, the troubled present, our anxious premonitions of the future, our neurotic conflicts and our inspired gropings toward the light.

Films are, in short, "a powerful touchstone into the unconscious."[6]

What do Indian films—and the popular addiction to them—suggest about the "psychic geography" of her millions? As has been noted, countless plots derive their impact from sexual concerns relating most often to young men and women. This is true of most film-making countries, but in India the concerns intersect with tensions over numerous other issues: the arranged marriage, caste, the status of women, the dowry, the joint family system, the place of the widow. All these issues involve age-old traditions that are challenged by pressures from the growth of industrialization, urbanization, and communication. The pressures cause uneasiness. Many in India, perhaps even more than elsewhere, have a feeling that the world they knew is slipping from under their feet, and that some-

[5] Bahadur, *The Context of Indian Film Culture,* p. 1.
[6] Greenberg, *The Movies on Your Mind,* pp. 3–4.

thing must be done to keep India Indian. While on one level they welcome the new, on another they cling to the old.

The tensions involved in this schism seem to underlie countless film plots, including those of song-and-dance films. Conflicts over marriage, caste, family duties, and the role of women provide the subject matter for innumerable films. But although the issues are raised, and the tensions exposed, the problems are seldom directly confronted. The industry has evolved story formulas with built-in mechanisms of evasion.

The chief evasive device is the use of the hero and the villain. If a story concerns the erosion of a joint family, the film may briefly touch on some of the factors undermining this ancient institution, but soon they are overshadowed by a villain who is held responsible for the family crisis. A hero confronts the villain, routs him, and the problem appears solved. Again and again, the hero-villain conflict washes out the social problem.

Script writers have instinctively gravitated toward story structures that allow audiences to have things both ways—to look at the truth and retreat to the myth, to enjoy the new but cling to the old. The strategy is especially evident in the handling of young romance. It has been tailored generally to the hopes and fears of the young urban male, and designed to ease his conflicts.

From tradition and custom, reinforced over the centuries by mythology and all the arts, the urban male has inherited the ideal of the father-worshiping, brother-worshiping, husband-worshiping Indian woman. She is the Savitri of Indian mythology, endlessly reincarnated in story heroines. Intellectually, he now rejects this ideal; emotionally, too, he is drawn to another image—the modern girl, free-thinking, accessible, sexually alluring, insistent on her own say in matters of romance—an international image that has a basis in fact and is also constantly underscored by the modern media, including advertising. The young Indian male would like such a girl, yet her independence threatens his ego, so well fortified by tradition. The Indian film lets him have his cake and eat it. A young film heroine, modern enough to choose her husband herself and often behaving with an exuberant informality, will after marriage fall at the feet of her husband and call him her "Kankanda Deivam"—the palpable god.[7] This female stereotype, fusing old and new, satisfies both the male libido and superego. The god needs someone to worship him, not just love him. Film provides her. Satish Bahadur calls her "Savitri in slacks." If such stereotypes dominate film stories, it is clear why actors—"film heroes" and "film heroines"—are likewise trapped in them.

Considering the adulation machinery surrounding the star—the seven hundred fan magazines, the fan clubs, the billboards dominating city streets and highways—it is perhaps strange that few feature films have been about movie stars, and those few have had little success. Guru Dutt's *Kaagaz Ke Phool* (Paper Flower), Shyam Benegal's *Bhumika* (The Role), and Satyajit Ray's *Nayak* (Hero), were all admired by critics; each is considered among its director's best work. Ray's *Nayak* is an especially incisive revelation of the phenomenon of stardom and the actor's problem of living with it.[8] But none of these films won mass audiences, perhaps because each sought to picture a human being behind the star façade—someone with fears and conflicts. This humanization of the star, disrupting the stereotype, was perhaps not what the fan was looking for.

The fan magazine treats many stars like mythological demigods who live on a highly physical and erotic plane, indulging in amours. The magazine speaks of the amours without any note of censure. They are apparently expected of stars, as of some mythological deities. In some fan magazines each number hints at newly developing liaisons and passionate affairs, and tries to pinpoint when they started. The stars are generally referred to by their first names. Compromising photos are a specialty of some of the magazines. Some feature jaunty question-and-answer sections about heroes, villains, and the film world. "Q. Why does the villain always stay fully dressed during the rape scene? A. Lack of sex education." Fan magazines appear in all major Indian languages as well as in English; the English-language magazines tend to be more scurrilous than the

[7] Krishnaswamy, "The Story of South Indian Cinema," *Illustrated Weekly of India,* January 10, 1965. The Tamil phrase has equivalents in other Indian languages.

[8] A notable feature of *Nayak* was that Ray persuaded a top-ranking "film hero," Uttam Kumar, to play the "hero," revealing a depth of acting talent seldom tapped in his formula films.

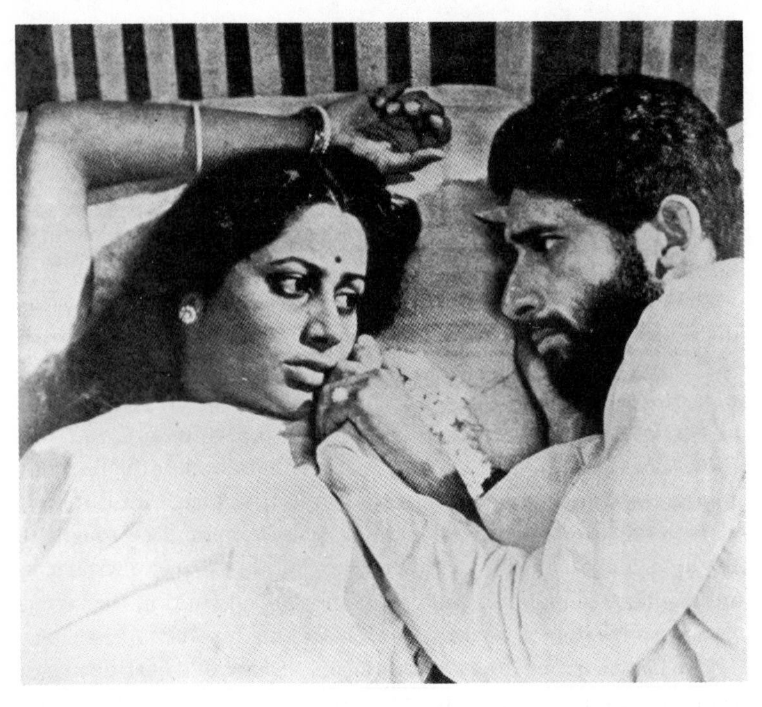

IN SHYAM BENEGAL'S *Bhumika* (THE ROLE), 1977: SMITA PATIL

others. Stars seldom sue fan magazines, possibly because careers do not seem to be damaged by them.

The Indian film world continues to be complex, chaotic, and rambunctious. It combines public and private enterprise in a uniquely Indian kaleidoscope. The government continues to dominate documentary and newsreel, while the feature film production remains the domain of fantastically fragmented private enterprise. But here, too, government injects itself at various points. It is a mixed economy with mixed results, ranging from the grossest to the finest in modern film. It is a wild gambling game, with large stakes lost or won overnight. New entrants continue to plunge in. A factor that brings them is black money, which a successful film can convert into respectable money. But the power and glitter of cinema also draw them. Some men, successful in other fields, move into film production for prestige.

While the smaller language areas have often been an innovative influence, the Hindi film continues to dominate, often penetrating other language areas. This has implications for the central government. It has been suggested that if Hindi ever wins acceptance as a national language, it will not be because of government efforts, but because of Hindi films.

The industry has few permanent institutions. A corporation formed for a specific film may dissolve and vanish without paying salaries. All talent is insecure. Some categories of workers have improved their lot during the 1960s and 1970s by way of a rising trade union movement. But others—especially extras—survive at precarious levels.[9]

Even the star is insecure. That appears to be why, at the top of his fame, he commits himself to dozens of films, allocating an hour here and an hour there, and keeping their producers all dancing attendance. The star's rationale: "Next year the offers may stop."

This combination of god-like status and insecurity has been vividly illustrated by Sivaji Ganesan. His standing among southern superstars was for a long time rivaled only by that of M. G. Ramachandran; for years a leading Madras theatre showed *only* films of Sivaji Ganesan. For years it seemed risky for a producer to produce a Tamil film *not* starring Sivaji Ganesan. Scores of producers, directors, actors, and technicians were continually dependent on his favorable decisions: his nod secured financial backing. During his brief appearances at a studio he would work with speed and precision, and charm everyone. Then he would be off again, leaving anxiety as to when he would return for another hour or so. Seldom has a substantial talent been used so recklessly—or so profitably. He amassed a fortune and was involved in well-organized and well-publicized charities. But he could view his own eminence objectively. Those who sought his favor, he said, had mixed feelings toward him. They wooed him but would also like to destroy him.

[9] Detailed in *Ai Anjane* (The Nobodies), 1973, an unusually candid documentary produced by P. V. Krishnamoorthy for Bombay television, and the object of indignant protests by the film industry.

MADRAS: SIVAJI GANESAN AND RANGA RAO IN *Annayin Anai* (A MOTHER'S COMMAND), 1958. PARAGON PICTURES

Asked if the dominance of the star was good for the industry, he said without hesitation that it was not.[10]

The government plays diverse roles toward the industry. Periodically it raids homes of stars, looking for money not accounted for. At festivals the ministers pose with stars. The government chastises and praises the industry. It gives awards. Ambivalence runs through all government-industry relations.

Government initiative has had mixed results. By requiring the showing of documentaries and newsreels, it has created a mechanism of great potential for a well-informed public, but results have consistently been disappointing. The Films Division monopoly control over "approved films" was never intended in the original legislation, but was created by the block-booking contract forced on theatres. This monopolistic system has had a sterilizing impact, and generated apathy. Independent producers have been brought into

[10] Interview, Ganesan.

the system under circumstances virtually guaranteed to negate their potential. Available talents have scarcely been tapped. And the elimination of all foreign documentaries from consideration as "approved films" cuts India off from much of value that might from time to time bring a breath of fresh air into the formula-ridden "approved film" schedule.

With the end of the Emergency, *autonomy* became the proclaimed goal—for the radio and television systems and ultimately for the Films Division. The "Verghese report"[11] which made this recommendation cited the example of the British Broadcasting Corporation, which has often shown independence of government administrations. But such independence requires more than a proclamation of autonomy; it requires a spirit nurtured by tradition. The development of such a tradition is surely to be hoped for.

Along with documentaries and newsreels, children's films have been a domain largely pre-empted by the government, through the establishment in 1955 of the Children's Film Society, a quasi-independent corporation provided with public funds to produce and distribute children's films. Here again the record has failed to justify the hopes stirred. Occasional private entrepreneurs, though entering this field at a considerable risk, have done somewhat better, offering such works as Satyajit Ray's *Sonar Kella* (The Golden Fortress) and Tapan Sinha's *Safed Hathi* (White Elephant), winner of a Best Children's Film award.

The government as catalyst, stimulating experimentation in the fiction film through the Film Finance Corporation, has been notably more successful. Its aid to new talent has had some significant results. In encouraging the film society movement through various mechanisms and regulations the government has also performed an important educational service, which it has further strengthened by the creation of the National Film Archive of India, while the Film and Television Institute of India, for the training of new talent, has unquestionable long-range value.

Amid these achievements, the government faces unsolved problems and challenges. They include the shortage of theatres, which seems to place limits on innovation and to bolster the proven for-

[11] *Akash Bharati: National Broadcast Trust,* Vols. I and II.

mulas. Producers argue that a large-scale cinema-building program would shift bargaining power away from exhibitors toward producers, and open the door to alternatives. The government decision to permit the nationalized insurance system to make loans for the building of rural cinemas was an important step. The Film Finance Corporation also resolved to move in this direction, but a lack of funds was a problem.

Into this situation the Kinematograph Renters' Society Ltd., representing major American producing and distributing companies, injected itself with a surprise suggestion. These companies had the problem of how to use blocked rupee funds in the Reserve Bank of India, earned in India by American films. The uses made of these funds require Indian government approval. The funds have at various times been used to finance the production of films or sequences in India. In 1978 the Society made a new and ingenious proposal. It offered an interest-free loan of Rs. 10,000,000 to the Film Finance Corporation to be used to stimulate the building of new Indian cinemas. Repayment of the loan would not begin until five years later (1983). Under arrangements subsequently approved by the government, the repaid sums would be convertible into dollars. Thus both the Kinematograph Renters' Society and the Film Finance Corporation saw value in the scheme. The FFC visualized a chain of small new theatres that would parallel "art theatres" in other countries.

Another government problem is that of Indian television, Doordarshan—in existence in a rudimentary form since 1959, but held back deliberately, and perhaps for good reason. In some countries, including the United States and Japan, the rapid rise of television brought devastating upheaval to the motion picture field, and many other economic and social reverberations. In many countries, the central role of the television sponsor was felt to have influenced programming, and to have fostered a culture obsessed with consumer goods. Indian government leaders recognize such a development as potentially dangerous for India, further widening the chasm between rich and poor, city and village. Yet the financing of an enlarged Doordarshan, without substantial advertising, offers a difficult fiscal challenge. The issue interlocks with India's ambivalent relations with multinational corporations, the mainstay of commercial television in many countries. Thus the problem of television has economic, social, and political ramifications far beyond questions of entertainment or even of education.

A third governmental problem is film censorship—a storm center since early motion picture days. Considering the anxieties stirred by rapid changes in Indian life, it is understandable that pressures for censorship remain strong. Meanwhile the industry continues to protest the strictness of censorship.

A favorite quotation of modern India is a statement of Mahatma Gandhi:

> I do not want my house to be walled in on all sides and my windows to be stuffed. I want the culture of all lands to be blown about my house as freely as possible, but I refuse to be blown off my feet by any of them.[12]

Many government officials have, in effect, revised this to make it read: "I do not want the culture of other lands to be blown about my house, lest it blow my people off their feet."

But the winds cannot be kept out. In a land of multiplying transistor radios—medium and short wave—they blow in on every channel from every quarter of the globe. They blow in with every foreign film in film society or festival or matinee. And these influences have, in fact, brought changes in censorship. General guidelines have replaced the more specific rules under which censorship panels operated in earlier years. More is left to interpretation, and interpretations of recent years indicate some liberalization, though with occasional setbacks. Visitors from abroad are at times surprised at the political thrust of some Indian films—a trend that apparently reflects changes in attitude within the screening panels, or within the appeals procedure.

Films are first screened by panels in Bombay, Madras, and Calcutta, any of which can certify a film for the whole of India. Unfavorable decisions can be taken to an appeals panel, and its decisions can be appealed to the Secretary of the Ministry of Information and Broadcasting—the final appeal level within the procedure. Reversals at this level have been obtained in some notable cases,

[12] Quoted in Nehru, *The Discovery of India,* p. 367.

including *Samskara* and *Garm Hawa*. Such reversals could not fail to influence subsequent deliberations of the regional panels.

In one unusual case, it was litigation that actuated the final reversal. A lawsuit was brought in 1968 by K. A. Abbas to challenge on constitutional grounds a censorship decision relating to his short film *A Tale of Four Cities*. This documentary paid brief tribute to accomplishments in housing, then stressed unfinished business by showing slums in the four major cities, including prostitute cages in Bombay. The Ministry offered Abbas an "adults only" certificate, provided cuts were made, but Abbas declined, asserting that an allegation of "pornography" was being used as a smokescreen for a political decision. He took his case to the Supreme Court, which appeared to agree with him. At this point the Ministry yielded, and granted a "universal" certificate. It was apparently the first successful use of litigation to challenge a censorship ban. Here, too, the event had the possibility of influencing subsequent censorship deliberations.

In the following years, except during the Emergency, censorship focused far less on ideas than on the curbing of sex and violence. These efforts, however well intentioned, tended to lead to tricky and fruitless games. Instead of the explicit, one got blatant sexual symbolism or coy suggestiveness, often more vulgar than explicitness. But such problems are inherent in the job of the censor, whose search for "objective" criteria inevitably leads to such traps. Perhaps the essential point is that censorship can curb this or that, but cannot create good films. It is therefore heartening that principal government efforts have shifted in other directions.

In a world in which outstanding films have come from both private and governmental film units, no one can any longer claim that the secrets belong exclusively to either. But films are made not by agencies but by people, and great films are made by those who know in their hearts what they must do, and have prepared intently for it, and are free to do it. Here the emphasis must surely lie.

India, struggling to rebuild its society, is moving into a position of leadership among nations. Film provides one of the voices through which it can speak to the world. If the films of Satyajit Ray and an increasing number of others have spoken in a meaningful way, it was not because these film makers were sheltered from alien influences, not because a censor kept them from going astray, not because they knew more technical tricks, but because they were in truly Indian fashion citizens of the world, who were free to speak to their fellow men as they knew they must, firm in the faith that "art wedded to truth must in the end have its reward."[13]

[13] Ray, "Problems of a Bengal Film Maker," in *International Film Annual,* No. 2, p. 53.

Reference

Production Statistics: Indian sound films, by language[1]

	Assamese	*Bengali*	*Gujarati*	*Hindi*	*Kannada*	*Malayalam*	*Marathi*	*Oriya*	*Punjabi*	*Tamil*	*Telugu*	*Other*[2]	*Totals*
1931	–	3	–	23	–	–	–	–	–	1	–	–	27
1932	–	4	2	61	–	–	8	–	–	4	3	1	83
1933	–	9	–	74	–	–	6	–	–	7	5	1	102
1934	–	10	1	121	2	–	11	–	–	14	3	2	164
1935	1	19	1	154	1	–	9	–	1	38	7	2	233
1936	–	19	3	135	1	–	6	1	1	38	12	1	217
1937	–	16	–	102	3	–	11	–	–	37	10	–	179
1938	–	19	–	88	–	1	14	–	1	39	10	–	172
1939	1	15	1	82	–	–	12	–	7	35	12	–	165
1940	–	16	1	86	–	1	10	–	7	36	14	–	171
1941	1	19	–	78	2	1	14	–	3	32	15	2	167
1942	–	18	–	98	2	–	13	–	5	23	12	2	173
1943	–	21	–	106	4	–	7	–	1	13	6	3	161
1944	–	14	–	85	–	–	5	–	2	14	6	1	127
1945	–	9	–	73	1	–	–	–	–	11	5	–	99
1946	–	15	1	155	–	–	2	–	1	16	9	–	199
1947	2	33	11	185	5	–	6	–	–	29	6	5	282
1948	–	37	27	147	2	1	7	1	1	32	7	1	263
1949	2	62	17	159	6	1	15	–	1	21	7	–	291
1950	–	42	13	114	1	6	19	2	5	19	18	2	241
1951	–	39	6	98	2	6	16	1	4	26	21	1	220
1952	–	43	2	102	1	11	17	–	–	32	25	–	233
1953	1	50	–	97	7	7	21	1	3	42	29	2	260
1954	1	48	–	108	11	9	18	1	3	38	28	3	268
1955	2	52	3	126	15	7	13	–	–	46	24	1	289

1956	3	54	3	123	14	5	13	2	–	51	27	1	296
1957	3	55	–	116	14	7	14	1	2	45	34	1	292
1958	2	45	1	114	11	4	16	–	1	61	36	3	294
1959	5	38	–	116	5	3	9	2	1	78	47	1	305
1960	–	36	2	119	12	6	15	5	4	64	53	2	318
1961	2	34	7	104	12	11	15	1	5	49	55	2	297
1962	–	35	5	91	16	15	22	7	6	60	48	3	308
1963	3	38	5	88	22	13	15	2	3	55	46	7	297
1964	1	38	2	98	18	19	17	3	8	44	41	14	303
1965	–	29	5	97	21	31	14	3	5	56	50	11	322
1966	2	30	2	99	21	32	12	2	4	62	37	4	307
1967	2	25	3	82	24	39	18	2	4	63	61	3	326
1968	1	29	3	73	34	36	17	3	2	63	70	3	334
1969	2	29	6	99	45	31	16	2	4	70	59	5	368
1970	3	34	5	102	37	44	19	–	2	75	72	3	396
1971	5	30	3	116	33	53	23	1	2	74	84	8	432
1972	7	25	4	132	20	47	12	1	3	77	73	10	411
1973	9	35	5	137	32	60	14	2	5	66	74	9	448
1974	3	35	7	132	30	54	11	2	4	79	69	6	432
1975	6	35	12	118	38	77	17	3	5	70	88	1	462
1976	5	32	29	106	45	84	10	6	10	81	93	6	507
1977	7	31	30	119	48	91	19	11	12	66	100	3	537
1978	6	37	32	122	54	123	15	15	8	105	94	8	619

[1] Courtesy B. V. Dharap, publisher, *Indian Films* annuals. The chart is based on censorship records, which vary somewhat from film industry lists.

[2] Other: Arabic, Avadhi, Bhojpuri, Burmese, Chhatisghari, Coorgi, Dogri, English, German, Haryani, Iranian, Kashmiri, Konkani, Magdhi, Maithili, Malay, Manipuri, Marwari, Nepalese, Persian, Pushtu, Rajasthani, Sindhi, Sinhalese, Thai, Tulu.

Chronology

1895 The Lumière brothers launch the Cinématographe on December 28 at the Grand Café in Paris.

1896 Lumière films are shown in Bombay at Watson's Hotel and the Novelty Theatre by Lumière *opérateur* Maurice Sestier en route to Australia.
Touring conjurors on all continents include short films in their "magic" acts.

1897 Maharashtrian photographer Harischandra Sakharam Bhatvedekar imports a motion picture camera and begins filming Indian "topicals."

1898 Short films shown at Indian theatres include such items as *Poona Races '98* and *Train Arriving at Bombay Station.*

1899 In Calcutta, Hiralal Sen films excerpts from plays at the Classic Theatre and shows them in conjunction with stage performances.

1900 Films imported into India average 60–75 feet (about 1 minute) but the length is increasing.

1901 Abdulally Esoofally begins South Asia tent-cinema travels in Singapore, eventually settling in India.
Films reaching India include *The Queen's Funeral Procession* and *Assassination of President McKinley* as well as scenes of the Boer War.

1902 In Calcutta, J. F. Madan launches regular "bioscope" showings in a tent on the Maidan.

1903 The Coronation Durbar for Edward VII is filmed by Bhatvadekar and many others.

1904 Films are coming into India from French, American, Italian, British, German, and Danish sources, with French films dominant.

1905 Topicals winning world distribution include fake scenes of the Russo-Japanese War and other "reconstructions."

1906 "Picture palaces" are becoming a feature of major cities in India as elsewhere.

1907 J. F. Madan of Calcutta begins to form a cinema chain, starting with the Elphinstone Picture Palace.

1908 "Films d'Art" featuring stage stars dominate imports from Europe, with "comics" also a growing attraction.

1909 The Madan cinema chain has grown to more than thirty cinemas.

1910 Dadasaheb Phalke, scholar and fine printer, sees the film *Life of*

Christ at the America-India Cinema in Bombay and determines to bring Krishna to the screen.

1911 Phalke travels to England for equipment and technical instruction.

1912 Phalke makes a film in time-lapse photography, to secure backing for his first feature.

1913 Phalke's feature-length *Rajah Harischandra* opens at the Coronation Cinema in Bombay and wins instant acclaim.
Phalke's second feature, *Bhasmasur Mohini* (The Legend of Basmasur), is the first Indian film to use an actress.

1914 With traveling film shows on the wane, Esoofally and partners take over the Alexandra Theatre in Bombay.

1915 As World War I curtails film production in France and England, the United States dominates all markets and introduces "block booking."

1916 Universal Pictures becomes the first American company to establish an agent in India.

1917 The Hindustan Film Company is formed by Phalke and financial backers.
To educate audiences, Phalke releases a documentary titled *How Films Are Made*.

1918 The Indian Cinematograph Act establishes a censorship and cinema-licensing system.

1919 Phalke's six-year-old daughter Mandakini plays the boy Krishna in *Kaliya Mardan* (The Slaying of the Serpent).

1920 Gandhi's call for non-violent non-cooperation brings ferment to the Indian film world.

1921 Dhiren Ganguly's *England Returned,* satirizing Indians imitating British ways, scores triumph in Calcutta.

1922 Ganguly forms the Lotus Film Company in Hyderabad, finds the Nizam of Hyderabad cooperative, and opens two cinemas.

1923 Cinemas in India now number more than 150, of which a third belong to the Madan chain.
J. F. Madan dies, and a Calcutta street is named after him.

1924 When one of Ganguly's cinemas exhibits *Razia Begum,* about a Hindu-Muslim romance, he is ordered to leave the domain of the Nizam of Hyderabad within twenty-four hours.

1925 In Bombay, Chandulal Shah produces *Gun Sundari* (Why Husbands Go Astray) and stimulates the rise of the "social film."
The Light of Asia, a German-Indian project organized by Himansu Rai, becomes the first international co-production involving India.

1926 Eighty-five percent of films shown in India are foreign—chiefly American—while 15 percent are Indian.

1927 An Indian Cinematograph Committee is appointed to study the cinema in India, and the feasibility of furthering "empire films" to counter American dominance.

1928 The six-man Indian Cinematograph Committee makes a report recommending support of *Indian* films, but the three British members issue a dissenting statement.

1929 The Prabhat Film Company is formed in Kolhapur.
Universal's *Melody of Love* inaugurates the sound-film era in India at Madan's Elphinstone Picture Palace, Calcutta.
Universal negotiates for the purchase of the Madan chain.
The Wall Street crash ends the Universal-Madan negotiations, while bringing widespread depression, unemployment, and unrest.

1930 Many Indian silent film companies collapse.
The British begin to ban all topicals on Gandhi's activities and demonstrations.
New Theatres is organized in Calcutta by B. N. Sircar.

1931 Ardeshir Irani's *Alam Ara* (Beauty of the World), in Hindi, becomes the first Indian sound feature.
J. J. Madan and brothers launch sound production activity, but then decide to liquidate the Madan film empire.

1932 The Motion Picture Society of India is formed.
Prabhat's *Ayòdhyecha Raja* (The King of Ayodhya), featuring Durga Khote, signals increasing acceptance of film by high-caste performers.

1933 Seventy-five Hindi films are produced, all with songs and dances.
The Prabhat Film Company moves to Poona and begins to develop a large studio complex.
Wadia Movietone is founded and begins to specialize in "stunt films."

1934 Himansu Rai's *Karma* (Fate), starring Devika Rani, opens in Bombay and leads to the formation of Bombay Talkies.

1935 P. C. Barua's *Devdas,* in a Bengali version and then a Hindi version starring Saigal, creates a sensation and popularizes the theme of the doomed hero.
The playback singer becomes a fixture in Indian film production.

1936 Modern Theatres Ltd. is formed in Salem by T. R. Sundaram.
Madras United Artistes Corporation is formed in Madras by K. Subrahmanyam.
The Tamil film *Balayogini* (Child Saint) features a Brahmin widow as a Brahmin widow, flouting ancient taboo.
The Bengal Motion Picture Association is established in Calcutta.
The Indian Motion Picture Producers Association is established in Bombay.

1937 Prabhat's *Sant Tukaram* becomes the first Indian film to win a Venice festival award.
Phalke makes his last film, *Gangavataran* (The Descent of Ganga).

1938 The South Indian Film Chamber of Commerce is established in Madras.

1939 Vauhini Pictures is formed in Madras by B. N. Reddi.

The Viceroy of India, Lord Linlithgow, declares India at war with Germany, and the Indian National Congress condemns the "undemocratic" action.

1940 The death of Himansu Rai.
A governmental Film Advisory Board is set up to encourage "war effort films."

1941 Minerva Movietone's *Sikander* (Alexander), depicting Alexander the Great's expedition to—and retreat from—India is banned from cantonment areas.
Independent production begins to undermine the major Indian studios.

1942 Portraits of Gandhi and Nehru, as well as Congress songs and symbols, are excised by censors from feature films.

1943 Showing of newsreels of *Indian News Parade* and documentaries of Information Films of India is made compulsory, with theatres paying a rental fee.

1944 "Black money" from war profits spurs production by "adventurers," inflating star salaries and hastening the decline of the studios.
D. G. Phalke dies, forgotten and destitute.

1945 *Dr. Kotnis Ki Amar Kahani* (The Journey of Dr. Kotnis) is approved as a "war effort" film.
The end of war brings renewed pressure for independence.

1946 *Indian News Parade* comes to an end, and Information Films of India is dissolved.

1947 Indian independence arouses vast hopes in the film industry but brings new taxes and crises.
The Calcutta Film Society is formed under the leadership of Satyajit Ray and Chidananda Das Gupta.
Shantaram's *Shakuntala* becomes the first Indian feature shown in a United States theatre.

1948 Communal strife, assassination of Gandhi, throw a pall over the film industry.
The Films Division is formed in the pattern of Information Films of India.
Uday Shankar's experimental *Kalpana* (Imagination) wins critical approval.
S. S. Vasan scores a nation-wide success with the superspectacle *Chandralekha*.

1949 In protest over taxes, the film industry stages an All-India Protest Day.
The government stirs hope with formation of a Film Enquiry Committee including film leaders.
Dharti Ke Lal (Children of the Earth) becomes the first Indian feature shown in Moscow.

1950 The rise of the Dravidian Forward Movement (DMK) is closely linked to Tamil film interests and the struggle against Hindi as a national language.
Jean Renoir makes *The River* in Bengal.

1951 The death of Barua.
Report of the Film Enquiry Committee urges creation of a Film Finance Corporation, an Institute of Film Art as a training school, and an Export Corporation to promote foreign markets.
The Film Federation of India is formed.
Pudovkin and other Soviet film leaders visit Indian film centers.
Films Division's *Jaipur* wins a first award for documentary at the Venice festival.

1952 The first International Film Festival of India is held.
Bombay Talkies ceases production.

1953 The Prabhat Film Company ceases production, virtually ending the big-studio era in India.

1954 Raj Kapoor's *Awara* (The Vagabond) is a smash hit in the Soviet Union.
Bimal Roy's *Do Bigha Zamin* (Two Acres of Land) wins awards at Cannes and Karlovy Vary.
National Awards for films are instituted.

1955 K. A. Abbas's *Munna* (The Lost Child), the first Hindi feature without songs or dances, wins kudos at the Edinburgh festival but fails financially in India.
Satyajit Ray completes his first film, *Pather Panchali* (Song of the Road).

1956 *Pather Panchali* wins the "best human document" award at Cannes.

1957 Satyajit Ray's *Aparajito* (The Unvanquished) wins the Golden Lion award at Venice.
Pardesi (The Traveler), produced by K. A. Abbas in collaboration with Mosfilm, becomes the first India–USSR co-production.

1958 M. G. Ramachandran's *Nadodi Mannan* (The Vagabond King) links heroic melodrama with Dravidian symbolism.

1959 *Kaagaz Ke Phool* (Paper Flower), starring its producer-director Guru Dutt, inaugurates wide-screen production in India with finesse but without success.
Television is begun experimentally by All India Radio with donated Phillips equipment.

1960 The Film Finance Corporation is formed.

1961 The Film Institute of India is established at Poona in the old Prabhat studios.
New Delhi hosts the second International Film Festival of India.

1962 Film societies win partial exemption from censorship.

1963 The fiftieth anniversary of feature production in India fosters new interest in Indian film history.
The Indian Motion Picture Export Corporation is formed.

1964 The National Film Archive of India is established at Poona.
Film society exemption from censorship is strengthened through arrangements negotiated by the Federation of Film Societies of India.

1965 New Delhi hosts the third International Film Festival of India, with additional screenings in Bombay, Calcutta, and Madras.

1966 The Dadasaheb Phalke awards are established to honor Indian film pioneers.

1967 Documentaries celebrating the twentieth anniversary of Indian independence include S. N. S. Sastry's *I Am Twenty* and S. Sukhdev's *India '67*.
Devika Rani receives the first Dadasaheb Phalke award.

1968 K. A. Abbas's *A Tale of Four Cities* precipitates a censorship struggle at the Supreme Court level.

1969 The fourth International Film Festival of India is held in New Delhi.
Mrinal Sen's *Bhuvan Shome,* financed by the Film Finance Corporation, wins multiple awards and commercial success.

1970 The Phalke birth centennial is celebrated as Bombay names a street after him.

1971 The Indian film industry becomes the world's most prolific in the production of theatrical features.

1972 Mrinal Sen's *Calcutta '71* injects the feature film forcefully into politics.
Film imports from the United States halted as a result of breakdown in negotiations with American film interests.

1973 M. S. Sathyu's *Garm Hawa* (Scorching Wind), funded by the Film Finance Corporation, wins multiple awards.

1974 *Ankur* (The Seedling), the first film by Shyam Benegal, wins major success and fosters the movement for "alternative" cinema.
Basu Chatterji's *Rajnigandha* (Tuberoses) scores major success for low-budget production.
The International Film Festival of India becomes an annual event.

1975 Loan of the American SITE satellite sets the stage for a one-year experiment in educational television by Doordarshan, the Indian television system.
Films from the United States again enter the Indian market.
Emergency proclamation brings tightened censorship and other governmental controls.

1976 The producers of the film *Kissa Kursi Ka* (Story of a Chair) accuse the government of illegally destroying the negative and all prints.

1977 Lifting of Emergency brings a gradual easing of governmental controls over the film industry.
Film star M. G. Ramachandran becomes Chief Minister of Tamilnadu but announces continuation of his film career.

1978 The Verghese committee proposes an autonomous status for Doordarshan and All India Radio.

1979 Satyajit Ray completes twenty-five years of feature film production. Indian production volume exceeds all previous records.

Interviews

Abbas, K. A., writer
Abraham, David, actor
Advani, L. K., public servant
Agarwal, G., distributor
Agte, B. Y., distributor
Ahmed, Mushir, public servant
Anand, Inder Raj, writer
Babu, Chitti, distributor
Bahadur, Satish, social scientist
Baji, A. R., public servant
Bakaya, M., executive
Banerjee, Jithen, technician
Banerjee, Kalyan, film society leader
Barua, Shyamalesh, technician
Benegal, Shyam, director
Bhanja, Manujendra, journalist
Bhasker, K. Dayanand, executive
Bhat, M. D., public servant
Bhatia, Vanraj, musicologist
Bhatiya, Mohan, producer
Bhattacharya, Basu, director
Billimoria, Fali, producer
Bode, Homi, journalist
Bose, Ajit, producer
Bose, Debaki, producer
Bose, Madhu, director
Chatterjee, M. D., distributor
Chatterji, Basu, director
Chettiyar, T. S. L. Chidambaram, financier
Choudhry, Salil, music director
Da Cunha, Uma, publicist
Dasarathy, B. S., public servant
Das Gupta, Chidananda, director
Das Gupta, Dhiren, producer
Das Gupta, Hari S., producer
Das Gupta, Prabir, distributor
Desai, Chimanlal, producer
Devarayalu, I. S. K., film society executive
Devi, Kanan, actress
Dewan, Karan, actor
Dharap, B. V., distributor
Dharker, Anil, public servant
Ganesan, Sivaji, actor
Ganguly, Dhiren, producer
Ghose, S., public servant
Ghosh, N. K., journalist
Gohar, actress
Gole, R. S., executive
Gopalan, S., public servant
Gupta, Anuva, actress
Irani, Ardeshir M., producer
Ishwar, R. V., executive
Ivory, James, director
Iyer, V. A. P., distributor
Jagirdar, Gajanan, actor
Jayakanthan, writer
Kannamba, actress
Kapoor, Ram, publicist
Kapoor, Shashi, actor
Karanjia, B. K., journalist
Karanth, B. V., director
Kariat, Ramu, director
Karnad, Girish, director
Keskar, B. V., public servant
Khan, Mehboob R., producer
Khandpur, K. L., public servant
Khemka, B. L., producer
Khote, Durga, actress
Khote, Tina, producer
Kothari, D. L., public servant
Krishnamoorthy, P. V., broadcaster
Krishnamurti, S., distributor
Krishnaswami, M. V., producer
Kulandaivelu, R., public servant

Madan, J. J., producer
Malik, Amita, journalist
Mani, Battling, actor
Mathur, H. B., film society executive
Mathur, J. C., public servant
Mathur, M. P., public servant
Mathur, P. S., public servant
Mathuram, T. A., actress
Mehra, Vinod, film society executive
Mehrotra, N. D., public servant
Mehta, C. C., writer
Meiyappan, A. V., producer
Menon, I. K., executive
Merchant, Ismail, producer
Mir, Ezra, producer
Mirza, Saeed, director
Mitra, Subrata, technician
Modi, Sohrab, producer
Mukherji, Subodh, executive
Munshi, K. M., lawyer
Munshi, Lilavati, publicist
Murari, Jagat, public servant
Murthy, N. V. K., public servant
Naidu, S. M. S., producer
Nair, M. T. Vasudevan, writer
Nair, P. K., archivist
Nath, Mahendra, executive
Natkarni, P. M., distributor
Padmanabhan, R., producer
Palekar, Amol, actor
Palekar, Chitra, actress
Parikh, Jagdish, executive
Parthasarathy, S., journalist
Pati, Promode, director
Patwardhan, N. J., art director
Pendarkhar, P. D., director
Phalke, Neelakanth, translator
Pillai, S. S., journalist
Pochee, E. A., distributor
Prabhu, P. V., distributor
Pramanick, D., executive
Pratap, K., public servant
Ragini, actress
Rajakumari, T. R., actress
Ramachandra, C., music director
Ramachandran, M. G., actor
Ramachandran, S., technician
Ramachandran, T. M., journalist
Ramadhyani, R. K., public servant
Raman, V., technician
Ramanathan, G., music director
Ramanujam, C. N., exhibitor
Ramnoth, T. V., journalist
Rao, Veeranki Rama, executive
Ray, Satyajit, director
Reddi, B. N., producer
Reddi, Gopala, public servant
Reddy, Pattabhi Rama, director
Roerich, Devaki Rani, actress
Roy, Bimal, producer
Sampath, E. V. K., public servant
Sanyal, Pahari, actor
Sanyal, Sudhirendra, journalist
Sarkar, Kobita, writer
Sastry, R. K., public servant
Sathe, V., writer
Sathyu, M. S., producer
Sen, Asit, director
Sen, Mrinal, producer
Seshadri, R. M., lawyer
Seton, Marie, writer
Seyne, Benoyendra, technician
Shah, Chandulal, producer
Shantaram, V., producer
Sharma, Kidar, producer
Sharma, Rajendra, director
Shaw, Alexander, executive
Shirur, R. M., distributor
Singh, Mala, producer
Sinha, Tapan, director
Sircar, B. N., producer
Srinivasan, C., executive
Subbalakshmi, S. D., actress
Subrahmanyam, K., producer
Subramaniam, C., public servant
Sukhdev, S., director
Sundaram, S. D., writer

Sundaram, T. R., producer
Thapar, Romesh, journalist
Vaidyanathan, K. S., distributor
Vasagam, S. K., journalist
Vasudev, Aruna, writer
Vasudevan, T. E., producer
Venkat, T. K., technician
Venkataraman, R., public servant
Venkatraman, K., financier
Wadia, J. B. H., producer

Bibliography

Abbas, Khwaja Ahmad. *I Am Not an Island: An Experiment in Autobiography*. New Delhi, Vikas, 1977.
——— "Mirror of India," *Theatre Arts*, February, 1948.
Agreement (printed form for contracts between central government and exhibitors). Bombay, Films Division, undated.
Akash Bharati: National Broadcast Trust: Report of the Working Group on Autonomy for Akashvani and Doordarshan. 2 Vols. New Delhi, Ministry of Information and Broadcasting, 1978.
Akashvani. New Delhi, Ministry of Information and Broadcasting, weekly.
American Film. Washington, American Film Institute, monthly.
Amrita. Calcutta, weekly.
Amrita Bazar Patrikà. Calcutta, daily.
Anderson, Joseph L., and Donald Richie. *The Japanese Film*. Rutland (Vt.) and Tokyo, Charles E. Tuttle, 1959.
Annual Report, 1955–56, 1956–57, etc. Madras, South Indian Film Chamber of Commerce, 1956, 1957, etc.
Asia. New York, monthly.
Asian Film Directory and Who's Who. *See* Doraiswamy, V., ed.
Audience Reaction to Films Screened in Villages: A Report on a Study of the Impact of Documentary Films in Tamil, Hindi and Bengali. New Delhi, Indian Institute of Mass Communication, 1969.
Azaad: Synopsis. Coimbatore, Pakshiraja Studio, 1955.
Bahadur, Satish, *The Context of Indian Film Culture*. Study Material Series No. 2, Poona, National Film Archive of India, 1978.
Balcon, Michael, Ernest Lindgren, Forsyth Hardy, and Roger Manvell. *Twenty Years of British Film, 1925–1945*. London, Falcon Press, 1947.
Bardèche, Maurice, and Robert Brasillach. *The History of Motion Pictures*. Trans. and ed. by Iris Barry. New York, W. W. Norton, 1938.
Barjatya, Tarachand. *A Handbook Detailing Some of the Problems of the Indian Film Industry*. Bombay, Rajshri Pictures Private Ltd., 1960.
Barnouw, Erik. *Documentary: A History of the Non-Fiction Film*. New York, Oxford University Press, 1974.
——— "The Magician and the Movies," *American Film*, April, May, 1978.
Basham, A. L. *The Wonder That Was India*. New York, Grove, 1959.
Basic Facts and Figures: International Statistics Relating to Education, Culture and Mass Communication. Paris, UNESCO, 1961.

Bauer, P. T. *Indian Economic Policy and Development.* London, George Allen & Unwin, 1961.

Bawa, Mohan. *Actors and Acting: 14 Candid Interviews.* Bombay, India Book House, 1978.

Baxter, John. *The Australian Cinema.* Sydney, Pacific Books, 1970.

Bhanja, Manujendra. "Crisis in the Cinema," *Hindustan Standard,* Puja Annual. Calcutta. 1961.

Bhanja, Manujendra, and N. K. G(hosh). "From *Jamai Sashti* to *Pather Panchali,*" in *Indian Talkie, 1931–56.* Bombay, Film Federation of India, 1956.

Bharat Jyoti. Bombay, weekly.

Bharucha, B. D., ed. *Indian Cinematograph Year Book, 1938.* Bombay, Motion Picture Society of India, 1938.

Bhatia, Vanraj. "Film Music," *Seminar,* music issue, December, 1961.

Bhatt, S. C. *Drama in Ancient India.* New Delhi, Amrit Book Company, 1961.

Bioscope. London, weekly. Discontinued.

Blitz. Bombay, weekly.

BMPA Journal. Calcutta, Bengal Motion Picture Association, monthly.

Boatwright, Howard. *Indian Classical Music and the Western Listener.* Bombay, Bharatiya Vidya Bhavan, 1960.

Bombay Cinema Regulations Act, 1953 (as modified up to October, 1960). Bombay, Government of Maharashtra Law and Judiciary Department, 1960.

Bombay Cinema Rules. Bombay, Government of Maharashtra, 1960.

Booch, Harish S. "Phalke, Father of Indian Motion Picture Industry," in *Indian Talkie, 1931–56.* Bombay, Film Federation of India, 1956.

Booch, S. H. *Film Industry in India.* New Delhi, India Information Services, undated (ca. 1953).

Bose, Debaki Kumar. "Films Must Mirror Life," in *Indian Talkie, 1931–56.* Bombay, Film Federation of India, 1956.

Bowers, Faubion. *Dance in India.* New York, Columbia University Press, 1953.

——— *Theatre in the East.* London, Thomas Nelson & Sons, Ltd., 1956.

Chakrabartty, Jagadish. "Bengal's Claim to Pioneership," *Dipali,* April 8, 1939.

Chatterji, Suniti Kumar. "Introduction," in *Indian Drama.* New Delhi, Ministry of Information and Broadcasting, 1959.

——— *Languages and the Linguistic Problem.* Oxford Pamphlets on Indian Affairs, No. 11. Madras, Oxford University Press, 1945.

Cine Advance. Calcutta, weekly.

Cinematograph Act, 1952, as modified up to June 1, 1959. New Delhi, Ministry of Law, 1959.

Conant, Michael. *Antitrust in the Motion Picture Industry.* Berkeley and Los Angeles, University of California Press, 1960.

Contemporary Indian Literature: A Symposium. New Delhi, Sahitya Akademi, 1959.
CSSEAS Review. Berkeley (Cal.), Center for South and Southeast Asia Studies, irregularly.
Das Gupta, Chidananda. "Reminiscence," *Seminar,* films issue, May, 1960.
Debi, Sabita. "Assamese Drama," in *Indian Drama.* New Delhi, Ministry of Information and Broadcasting, 1959.
Desai, Chimanlal. "Sagar Was Ten Years Ahead," in *Indian Talkie, 1931–56.* Bombay, Film Federation of India, 1956.
Desai, Nanabhai R. "Overseas Markets for Indian Films," in *Indian Cinematograph Year Book, 1938,* ed. B. D. Bharucha. Bombay, Motion Picture Society of India, 1938.
Deshpande, G. T. "Sanskrit Drama," in *Indian Drama.* New Delhi, Ministry of Information and Broadcasting, 1959.
Deslandes, Jacques, and Jacques Richard. *Histoire Comparée du Cinéma.* Vol. II. Casterman, n.p., 1968.
Developing Mass Media in Asia. Paris, UNESCO, 1960.
Devi, Kanan. "Rise of the Star System," in *Indian Talkie, 1931–56.* Bombay, Film Federation of India, 1956.
Dipali. Calcutta, weekly. Discontinued, 1950.
Doraiswamy, V., ed. *Asian Film Directory and Who's Who, 1956.* Bombay, 1956.
Era. London, weekly. Discontinued, 1939.
Evidence, Indian Cinematograph Committee, 1927–1928. 4 Vols. Calcutta, Government of India Central Publications Branch, 1928.
Far Eastern Survey. New York, Institute of Pacific Relations, biweekly.
Fathelal, S. "Prabhat Was a Training School," in *Indian Talkie, 1931–56.* Bombay, Film Federation of India, 1956.
Fazalbhoy, Y. A. *The Indian Film: A Review.* Bombay, Bombay Radio Press, 1939.
Fielden, Lionel. *The Natural Bent.* London, André Deutsch, 1960.
Film Comment. New York, monthly.
Film Daily. New York, daily.
Film in India. New Delhi, Ministry of Information and Broadcasting, undated (ca. 1957).
Film Institute of India: A Prospectus. New Delhi, Ministry of Information and Broadcasting, 1961.
Film Seminar Report. See Roy, R. M., ed.
Filmfare. Bombay, weekly.
Filmindia. Bombay, monthly. Succeded by *Mother India,* 1961.
Filmland. Calcutta, weekly. Discontinued.
Films Division. New Delhi, Ministry of Information & Broadcasting, 1961.
Flash. Madras, biweekly.
Garga, B. D. "Beyond Our Frontiers," *Filmfare,* November 3, 1961.
——— "A Critical Survey," *Marg,* documentary film issue, June, 1960.

——— "Historical Survey," *Seminar,* films issue, May, 1960.
Gargi, Balwant. "Role of Tradition," *Seminar,* stage issue, April, 1962.
——— *Theatre in India.* New York, Theatre Arts Books, 1962.
Gaur, Madan. *Other Side of the Coin: An Intimate Study of the Indian Film Industry.* Bombay, Trimurti Prakashan, 1973.
Gopalakrishnan, V. S. "Four Decades of Tamil Films," *Filmfare,* April 20, 1962.
Greenberg, Harvey R. *The Movies On Your Mind.* New York, Dutton, 1975.
Gunther, John. *Inside Russia Today.* New York, Harper, 1958.
Gupta, Sehdev Kumar. *Address, Symposium on Indian Cinema.* 33rd International Film Festival, Venice.
"Half a Century in Exhibition Line: Shri Abdulally Recalls Bioscope Days," in *Indian Talkie, 1931–56.* Bombay, Film Federation of India, 1956.
Handbook of the Indian Film Industry. Bombay, Motion Picture Society of India, 1949.
Hendricks, Gordon. *Origins of the American Film.* New York, Arno, 1972.
Hertz, Carl. *A Modern Mystery Merchant.* London, Hutchinson, 1924.
Hindu. Madras, daily.
Hindustan Standard. Calcutta, daily.
Holmes, Winifred. *Orient: A Survey of Films Produced in Countries of Arab and Asian Culture.* London, British Film Institute, 1959.
Illustrated Weekly of India. Bombay, weekly.
In the Court of the Industrial Tribunal: Labour Dispute between the Workers and Management of Cinema Talkies in Madras City. Madras, Government Press, 1948.
India, 1960, 1961, etc. New Delhi, Ministry of Information and Broadcasting, 1960, 1961, etc.
Indian Cinematograph Committee, 1927–28. For Evidence, Vols. I–IV, *see Evidence.* For Report, *see Report of the Indian Cinematograph Committee, 1927–1928.*
Indian Cinematograph Year Book, 1938. See Bharucha, B. D., ed.
Indian Documentary. Bombay, Paul Zils, monthly. Discontinued, 1959.
Indian Drama. New Delhi, Ministry of Information and Broadcasting, 1959.
Indian Express. Bombay, New Delhi, etc., daily.
Indian Films, 1972, 1973, etc. Annual series. Poona, B. V. Dharap, 1973, 1974, etc.
Indian Motion Picture Almanac and Who's Who, 1953. Bombay, Film Federation of India, 1953.
Indian Talkie, 1931–56 (Silver Jubilee Souvenir). Bombay, Film Federation of India, 1956.
International Film Annual, No. 2. *See* Whitebait, William, ed.

International Motion Picture Almanac. New York, Quigley Publications, annual.

Jain, Rikhab Dass. *The Economic Aspects of the Film Industry in India.* Delhi, Atma Ram & Sons, 1960.

Journal of the Film Chamber. Madras, South Indian Film Chamber of Commerce, monthly.

Journal of the Film Industry. Bombay, Indian Motion Picture Producers Association, monthly.

Journal of the Motion Picture Society of India. Bombay, Motion Picture Society of India, monthly. Discontinued, 1937.

Kabir, Humayun. *Britain and India.* New Delhi, Indian Council for Cultural Relations, 1960.

Keene, Ralph. "Cast and Caste: Some Problems of Making a Documentary in Assam," in *Penguin Film Review,* No. 5. London and New York, Penguin Books, 1947.

Khanna, Satti. "Reviews," *CSSEAS Review,* 1975.

Khosla, G. D. *Pornography and Censorship in India.* New Delhi, Indian Book Company, 1976.

Knight, Arthur. *The Liveliest Art: A Panoramic History of the Movies.* New York, New American Library, 1978.

Kracauer, Siegfried. *From Caligari to Hitler: A Psychological Study of the German Film.* Princeton, Princeton University Press, 1947.

Kripalani, J. B. "Deep Roots," *Seminar,* corruption issue, April 1960.

Krishnaswamy, S. "Critics and Yardsticks," *Illustrated Weekly of India,* May 1, 1966.

——— "The Story of South Indian Cinema," *Illustrated Weekly of India,* January 10, 1965.

Leyda, Jay. *Dianying: Electric Shadows. An Account of Films and the Film Audience in China.* Cambridge (Mass.), MIT Press, 1972.

——— *Films Beget Films.* New York, Hill and Wang, 1964.

——— *Kino: A History of the Russian and Soviet Film.* London, George Allen & Unwin, 1960.

Lok Sabha Debates. New Delhi, Lok Sabha Secretariat, irregularly.

Low, Rachael. *The History of the British Film, 1906–1914.* London, George Allen & Unwin, 1949.

——— *The History of the British Film, 1914–1918.* London, George Allen & Unwin, 1950.

Low, Rachael, and Roger Manvell. *The History of the British Film, 1896–1906.* London, George Allen & Unwin, 1948.

Madras Filmdiary, 1957, 1958, etc. Madras, Veeranki Rama Rao, 1957, 1958, etc.

Mahmood, Hameeduddin. *The Kaleidoscope of Indian Cinema.* New Delhi, Affiliated East-West Press, 1975.

Mail. Madras, daily.

Malik, Amita. "Padma Shri Devika Rani," *Filmfare,* March 14, 1958.
Manchester Guardian, later the *Guardian* (England), daily.
Mani, A. D. "The Indian Press Today," *Far Eastern Survey,* July 2, 1952.
Marg: A Magazine of the Arts. Bombay, monthly.
Mass Media in India, 1978. New Delhi, Ministry of Information and Broadcasting, 1978.
Mehta, Ved. *The New India.* New York, Penguin, 1978.
Mohan, Jag. "Panorama of the Private Sector in Indian Short Film Industry," *Marg,* June, 1960.
Montage, Nos. 5 and 6. Special Issue on Satyajit Ray. Bombay, Anandam Film Society, 1966.
Monthly Statistics of the Foreign Trade of India. Calcutta, Department of Commercial Intelligence and Statistics, monthly.
Motion Picture Year Book of Asia, 1956, 1957. Tokyo, Far East Film News, 1956, 1957.
Motion Pictures Abroad: India. Washington, U.S. Department of Commerce, 1961.
National Herald. Lucknow, daily.
Natyasastra: A Treatise on Hindu Dramaturgy and Histrionics. Ascribed to Bharata Muni. Vol. 1. Trans. by Manomohan Ghosh. Calcutta, Royal Asiatic Society of Bengal, 1950.
Nehru, Jawaharlal. *The Discovery of India.* London, Meridian Books. 1956.
——— *India Today and Tomorrow.* New Delhi, Indian Council for Cultural Relations, 1960.
——— *Toward Freedom.* New York, John Day, 1942.
News Chronicle. London, daily. Incorporated 1960 with *Daily Mail.*
Nichols, Beverley. *Verdict on India.* New York, Harcourt Brace, 1944.
Panikkar, K. M. *Common Sense about India.* London, Victor Gollancz, 1960.
Parrain, P. *Regards sur le Cinéma Indien.* Paris, Éditions du Cerf, 1969.
Paul, Robert William. B. K. S. lecture, February 3, 1936. British Kinematograph Society.
Penguin Film Review, No. 5. London, Penguin, 1947.
Phalke: Commemoration Souvenir. Bombay, Phalke Centenary Celebrations Committee, 1970.
Phalke, Suresh. "The Film Industry and Phalke," *Hindustan Standard,* February 26, 1961.
Rahim, N. K. "The Film in India," in *Penguin Film Review,* No. 5. London and New York, Penguin Books, 1947.
Rai, Himansu. "Light of Asia: Budha's Early Life Shown in a Pioneer Enterprise in Indian Film Making," *Asia,* September, 1926.
Ramadhyani, R. K. *Essays and Addresses.* Delhi, Atma Ram & Sons, 1961.
Ramsaye, Terry. *A Million and One Nights.* New York, Simon and Schuster, 1926.

Rangaswami, S., ed. *Who Is Who in Indian Filmland.* Madras, Happy India Office, 1933.

Rangoonwalla, Firoze. *Seventy-Five Years of Indian Cinema.* New Delhi, India Library, 1975.

Ray, Satyajit. *Our Films, Their Films.* Calcutta, Orient Longmans, 1976.

——— "Problems of a Bengal Film Maker," in *International Film Annual,* No. 2. Ed. by William Whitebait. New York, Doubleday, 1958.

Report, 1954–55, 1955–56, etc. New Delhi, Ministry of Information and Broadcasting, 1955, 1956, etc.

Report on an Enquiry into the Conditions of Labour in the Cinema Industry in Bombay State. Bombay, Government of Bombay State, 1956.

Report of the Film Enquiry Committee. New Delhi, Government of India Press, 1951.

Report of the Indian Cinematograph Committee, 1927–28. Madras, Government Press, 1928.

Report of the Society for the Prevention of Unhealthy Trends in Motion Pictures. No. 1, to November 15, 1959. No. 2, to June 20, 1961. Bombay, 1959, 1961.

Robinson, David. "Sixty Years of Japanese Cinema," in *International Film Annual,* No. 2. Ed. by William Whitebait, New York, Doubleday, 1958.

Rotha, Paul, and Richard Griffith. *The Film Till Now.* London, Vision Press and Mayflower Publishing Company, 1960.

Rotha, Paul, Sinclair Road, and Richard Griffith. *Documentary Film.* London, Faber and Faber, 1952.

Roy, R. M., ed. *Film Seminar Report.* New Delhi, Sangeet Natak Akadami, 1955.

Sadoul, Georges. *Histoire du Cinéma Mondial: Des Origines à Nos Jours.* Paris, Flammarion, 1949.

——— *Louis Lumière.* Paris, Seghers, 1964.

Sarkar, Kobita. *Indian Cinema Today.* New Delhi, Sterling, 1975.

Sastri, K. A. Nilakanta. *A History of South India.* Madras, Oxford University Press, 1958.

Scotsman. Edinburgh, daily.

Screen. Bombay, Express Newspapers, weekly.

Screen Year Book and Who's Who, 1956. Bombay, Express Newspapers, 1956.

Seminar. New Delhi, monthly.

Sen, Mrinal. *Views on Cinema.* Calcutta, Ishan, 1977.

Sen, Prabodh. "Bengali Drama and Stage," in *Indian Drama.* New Delhi, Ministry of Information and Broadcasting, 1959.

Sen, Sukumar. *History of Bengali Literature.* New Delhi, Sahitya Akademi, 1960.

Seton, Marie. *Portrait of a Director: Satyajit Ray.* New Delhi, Vikas, 1976.
——— "Pursuit for Reality: The Development of Documentary Films," *Marg,* June, 1960.
——— "The Year in Films," *Shankar's Weekly,* May 20, 1962.
Shah, Panna. *The Indian Film.* Bombay, Motion Picture Society of India, 1950.
Shankar's Weekly. New Delhi, weekly.
South Indian Film Chamber of Commerce: 20 Years of Service, 1939–1959. Madras, South Indian Film Chamber of Commerce, 1959.
Star. London, daily. Succeeded 1960 by *Evening News and Star.*
State Awards for Films, 1955, 1956, etc. New Delhi, Ministry of Information and Broadcasting, 1955, 1956, etc.
Statesman. Calcutta and New Delhi, daily.
Theatre Arts. New York, monthly.
Times of India, Bombay and New Delhi, daily.
Toynbee, Arnold. *One World and India.* New Delhi, Indian Council for Cultural Relations, 1960.
Variety. New York, weekly.
Vasan, S. S. "Pageants for our Peasants," in *Indian Talkie, 1931–56.* Bombay, Film Federation of India, 1956.
Vasudev, Aruna. *Liberty and Licence in the Indian Cinema.* New Delhi, Vikas, 1978.
Vyas, Narottam. "Writers Were Better Respected," in *Indian Talkie, 1931–56.* Bombay, Film Federation of India, 1956.
Wallbank, T. Walter. *A Short History of India and Pakistan.* New York, New American Library, 1958.
Whitebait, William, ed. *International Film Annual,* No. 2, New York, Doubleday, 1958.
Who Is Who in Indian Filmland. See Rangaswami, S., ed.
Wood, Robin. *The Apu Trilogy.* New York, Praeger, 1971.
World Communication: Press, Radio, Film, Television. Paris, UNESCO, 1950, 1951, etc.
World Screen. Rome, International Film and Television Council, monthly.

Index